ADVANCES IN THE PSYCHOBIOLOGY OF SLEEP AND CIRCADIAN RHYTHMS

Advances in the Psychobiology of Sleep and Circadian Rhythms features international experts from the fields of psychobiology, sleep research and chronobiology to address and review cutting-edge scientific literature concerning recent advances in the psychobiology of sleep, sleep disorders, such as sleep apnoea and insomnia, and circadian rhythms, across the lifespan.

In this illuminating volume, Melinda L. Jackson and Sean P.A. Drummond bring together leading international researchers to review cross-cutting issues in the field, including sleep and pain, sleep and dementia risk, and sleep issues in paediatric populations as well as the interaction between sleep and health conditions in different populations. The chapters offer coverage of the major explanatory models which underpin the empirical work as well as a discussion of the relevant theoretical and conceptual models on issues arising with specific psychiatric and medical disorders, including depression, dementia, posttraumatic stress disorder and pain. They also address new research in the area of chronobiology, and circadian impacts on health and diseases. The chapters also discuss important methodological and ethical issues arising in research and include sections addressing implications for public policy and practitioner interventions in the context of different social and cultural environments.

This volume will be a crucial resource for professionals, practitioners and researchers engaged in the field as well as for postgraduate and upper-level undergraduate students undertaking research in areas related to psychobiology, neuropsychology, health psychology and other disciplines such as biology, physiology and psychopharmacology.

Melinda L. Jackson is Associate Professor of Psychology in the School of Psychological Sciences and the Turner Institute for Brain and Mental Health, Monash University. She is also Honorary Research Fellow at the Institute for Breathing and Sleep, Austin Health.

Sean P.A. Drummond is Professor of Clinical Neuroscience at Monash School of Psychological Sciences and the Turner Institute for Brain and Mental Health. He currently also serves as the Director of Research Programs and Infrastructure.

ADVANCES IN THE PSYCHOBIOLOGY OF SLEEP AND CIRCADIAN RHYTHMS

Edited by Melinda L. Jackson and Sean P.A. Drummond

Routledge
Taylor & Francis Group

LONDON AND NEW YORK

Cover image: © GettyImages/metamorworks

First published 2024
by Routledge
4 Park Square, Milton Park, Abingdon, Oxon OX14 4RN

and by Routledge
605 Third Avenue, New York, NY 10158

Routledge is an imprint of the Taylor & Francis Group, an informa business

British Library Cataloguing-in-Publication Data
A catalogue record for this book is available from the British Library

ISBN: 978-1-032-28464-4 (hbk)
ISBN: 978-1-032-28459-0 (pbk)
ISBN: 978-1-003-29696-6 (ebk)

DOI: 10.4324/9781003296966

Typeset in Galliard
by Apex CoVantage, LLC

(From Melinda L. Jackson) To Finn; for teaching me the true meaning of a good nights' sleep

CONTENTS

CONTRIBUTORS

Kira Abirgas San Diego VA Healthcare System and University of California, San Diego USA
https://orcid.org/0000-0002-7714-8439

Ellemarije Altena Université de Bordeaux, CNRS UMR 5287, INCIA, Bordeaux France
https://orcid.org/0000-0002-8882-7963

Thomas Andrillon Sorbonne Université, Institut du Cerveau; Paris Brain Institute; ICM, Inserm, CNRS, AP-HP, Hôpital de la Pitié Salpêtrière France
https://orcid.org/0000-0003-2794-8494

Bryan S. Baxter Department of Psychiatry, Massachusetts General Hospital, Harvard Medical School USA
https://orcid.org/0000-0003-3056-0204

Thomas Bilterys Pain in Motion International Research Group, Department of Physiotherapy, Human Physiology and Anatomy (KIMA), Faculty of Physical Education & Physiotherapy, Vrije Universiteit Brussel, Belgium; Department of Psychology, University of Warwick United Kingdom; Institute of Advanced Study, University of Warwick, UK
https://orcid.org/0000-0002-5673-7314

Ryan Bottary Institute for Graduate Clinical Psychology, Widener University, Chester, PA USA
https://orcid.org/0000-0003-4495-5354

Romola S. Bucks University of Western Australia Australia
https://orcid.org/0000-0002-4207-4724

Peter J. Colvonen San Diego VA Healthcare System and University of California, San Diego USA
https://orcid.org/0000-0003-0222-8781

Tony J. Cunningham Center for Sleep and Cognition, Department of Psychiatry, Beth Israel Deaconess Medical Center & Division of Sleep Medicine, Harvard Medical School USA
https://orcid.org/0000-0003-4187-9640

Dan Denis Department of Psychology, University of York United Kingdom
https://orcid.org/0000-0003-3740-7587

Sean P.A. Drummond Turner Institute for Brain and Mental Health, School of Psychological Sciences, Monash University Australia

Symielle A. Gaston Epidemiology Branch, National Institute of Environmental Health Sciences, National Institutes of Health, Department of Health and Human Services USA
https://orcid.org/ORCID 0000-0001-9495-1592

Suzanne B. Gorovoy Sleep and Health Research Program, Department of Psychiatry, University of Arizona College of Medicine USA
https://orcid.org/0000-0002-7043-6489

Michael A. Grandner Sleep and Health Research Program, Department of Psychiatry, University of Arizona College of Medicine USA
https://orcid.org/0000-0002-4626-754X

Camilla Hoyos CIRUS Centre for Sleep and Chronobiology, Woolcock Institute of Medical Research Faculty of Medicine, Health and Human Sciences, Macquarie University Australia
https://orcid.org/0000-0002-6543-4016

Chandra L. Jackson Epidemiology Branch, National Institute of Environmental Health Sciences, National Institutes of Health, Department of Health and Human Services
Division of Intramural Research, National Institute on Minority Health and Health Disparities, National Institutes of Health, Department of Health and Human Services USA
https://orcid.org/ORCID 0000-0002-0915-8272

Melinda L. Jackson Turner Institute for Brain and Mental Health, School of Psychological Sciences, Monash University Australia
https://orcid.org/0000-0003-4976-8101

Aaron Lam Healthy Brain Ageing Program, Brain and Mind Centre, University of Sydney Australia
https://orcid.org/0000-0001-6087-3565

Bastien Lechat Flinders Health and Medical Research Institute: Sleep Health, Flinders University Australia
https://orcid.org/0000-0003-0760-0714

Gina M. Mason EP Bradley Hospital Sleep Research Laboratory, Department of Psychiatry and Human Behavior, Alpert Medical School of Brown University USA
https://orcid.org/0000-0003-0306-765X

Jade M. Murray Turner Institute for Brain and Mental Health, School of Psychological Sciences, Monash University Australia
https://orcid.org/0000-0001-9377-5750

Sharon L. Naismith Healthy Brain Ageing Program, Brain and Mind Centre, University of Sydney Australia
https://orcid.org/0000-0001-9076-2778

Jo Nijs Pain in Motion International Research Group, Department of Physiotherapy, Human Physiology and Anatomy (KIMA), Faculty of Physical Education & Physiotherapy, Vrije Universiteit Brussel, Brussels
    Department of Physical Medicine and Physiotherapy, University Hospital Brussels, Brussels Belgium
https://orcid.org/0000-0002-4976-6563

Delphine Oudiette Sorbonne Université, Institut du Cerveau; Paris Brain Institute; ICM, Inserm, CNRS, AP-HP, Hôpital de la Pitié Salpêtrière France
https://orcid.org/0000-0002-6598-4130

Craig Phillips CIRUS Centre for Sleep and Chronobiology, Woolcock Institute of Medical Research Faculty of Medicine, Health and Human Sciences, Macquarie University Australia
https://orcid.org/0000-0002-9126-6757

Sara Rama San Francisco VA Health Care System & Department of Psychiatry, University of California, San Francisco USA

Elizabeth F. Rasmussen Sleep and Health Research Program, Department of Psychiatry, University of Arizona College of Medicine USA
https://orcid.org/0000-0001-5542-7245

Genevieve Rayner Melbourne School of Psychological Sciences, The University of Melbourne Australia
https://orcid.org/0000-0002-0747-3877

Ivana Rosenzweig Sleep and Brain Plasticity Centre, King's College London United Kingdom
https://orcid.org/0000-0003-2152-9694

Jared M. Saletin EP Bradley Hospital Sleep Research Laboratory, Department of Psychiatry and Human Behavior, Alpert Medical School of Brown University USA
https://orcid.org/0000-0002-8547-0161

Hannah Scott Flinders Health and Medical Research Institute: Sleep Health, Flinders University Australia
https://orcid.org/0000-0002-0707-8068

Lin Shen Turner Institute for Brain and Mental Health, School of Psychological Sciences, Monash University Australia
https://orcid.org/0000-0001-8977-8625

Rupsha Singh Division of Intramural Research, National Institute on Minority Health and Health Disparities, National Institutes of Health, Department of Health and Human Services USA

Tracey L. Sletten Turner Institute for Brain and Mental Health, School of Psychological Sciences, Monash University Australia
https://orcid.org/0000-0002-0005-7838

Laura D. Straus San Francisco VA Health Care System & Department of Psychiatry, University of California, San Francisco USA
https://orcid.org/0000-0002-7859-8108

Nicole Tang Department of Psychology, University of Warwick United Kingdom
https://orcid.org/0000-0001-7836-9965

Prerna Varma Turner Institute for Brain and Mental Health, School of Psychological Sciences, Monash University Australia
https://orcid.org/0000-0001-5408-1625

FOREWORD

This is the second volume in the series published by Routledge entitled *Current Issues in Psychobiology* and follows the earlier publication of the volume entitled *Psychobiological Issues in Substance Use and Misuse*. As the series editor I would like to thank the editors of the current volume, Associate Professor Melinda L. Jackson and Professor Sean P.A. Drummond, both of Monash University in Melbourne, for their conscientious work in making its publication possible. The COVID-19 pandemic has disrupted the workload of academics around the world, so it has become increasingly difficult to recruit busy researchers to a project such as this, so the high scientific quality of the chapters contained in this book says much about the dedication of Melinda and Sean to the task in hand.

The creation of this series follows in the wake of the successful publication in 2018 of the *Routledge International Handbook of Psychobiology*, which I also had the privilege to edit. Psychobiological research covers a broad range of areas, and the aim of each volume in the series is to present a collection of cutting-edge reflective reviews, addressing important issues in a given area. Each chapter is written by one or more active researchers in that particular area, and each volume is edited by one or more experienced senior researchers in that area. Our target audience for these volumes spans from undergraduates seeking to specialise in their studies to experienced professional academics seeking to stay up to date with the literature in a particular research area. The series is also relevant to practitioners in various fields who want to develop the evidence-based nature of their practice.

The present volume, *Advances in the Psychobiology of Sleep and Circadian Rhythms*, comprises 12 chapters, representing the input of 36 authors in total, based in 24 institutions ranged across three continents. Each of the chapters was written to meet the criteria, stated above, of being cutting-edge reviews of ongoing research questions in the study of sleep whilst including elements of reflection intended to facilitate further developments in research. This volume would obviously not exist without the generous efforts of the chapter authors, all of whom are thanked here for their contributions.

Future volumes being planned for the *Current Issues in Psychobiology* will focus upon areas such as sport and physical activity, nutrition, stress and development across the lifespan. In the meantime, it is hoped that this present volume will meet your needs as a reader for an authoritative review of important questions in *Advances in the Psychobiology of Sleep and Circadian Rhythms*.

Philip N. Murphy, PhD
Professor of Psychology,
Edge Hill University, UK.
Series editor: *Current Issues in Psychobiology*

PREFACE

It has been known since ancient times that good sleep is vital to daytime function. However, it was not until the 1890s that the first peer-reviewed publication on sleep and cognition appeared,[1] and it was not until the 1950s that sleep could be objectively observed and studied through the measurement of brain electrical activity.[2] Over the past 70 years, since the discovery of REM sleep, a tremendous amount of research has sought to elucidate the roles and functions of sleep.

Today, we know sleep touches literally every system in the body, and it is especially critical for optimal brain function. At the same time, physical and mental health often impact sleep. Interestingly, while these relationships are clearly bidirectional, the influence of sleep on health is often stronger than the other way around. Moreover, sleep has been linked with a broad array of processes beyond health, ranging from learning and memory to emotional regulation and mood.[3] These discoveries have been made through the use of a variety of methodological approaches and techniques, spanning neuroimaging, electrophysiological and behavioural, to reveal the complex relationships between sleep and functioning of the awake brain.

As outlined above, the fields of sleep and chronobiology are a productive and growing area of research. These research fields are particularly relevant to psychobiological researchers given the overlapping neurobiological underpinnings and high comorbidity between sleep and circadian rhythms, with mental health and medical conditions. As such, the aim of this book *Advances in the Psychobiology of Sleep and Circadian Rhythms* is to provide the reader with a collection of cutting-edge reviews of the scientific literature concerning recent advances in the psychobiology of sleep across the lifespan. It is our goal to summarise just of some of the exciting new areas of research currently under investigation in the field.

The first two chapters focus on two fundamental aspects of normal, healthy sleep. Andrillon and Delphine discuss new advances in our understanding of the sleeping brain from the perspective of local regulation of sleep. They further discuss how this fascinating concept helps us to understand atypical phenomena such as dreaming in NREM sleep and mind wandering. Chapter 2, by Bottary and colleagues, discusses the well-established links between sleep and memory consolidation, describing new insights into the neurobiological underpinnings of these 'offline' processes.

We then turn to clinical perspectives on sleep and psychobiology over the next four chapters. Insomnia is the most common sleep disorder worldwide, affecting around 10% of the population. The chapter by Altena outlines the current knowledge regarding the neurobiology of insomnia and both daytime and night-time factors known to impact sleep onset and maintenance. Almost all psychiatric and neurological conditions have sleep problems as part of the diagnostic criteria, and indeed up to 90% of individuals with these conditions report either insomnia or hypersomnolence. The chapters by Straus et al. and Jackson et al. focus on two common psychiatric conditions – PTSD and depression – outlining how sleep and sleep disorders interact with these conditions, and influence treatment outcomes.

Dementia is an increasingly prevalent neurological condition. While a number of medical and lifestyle risk factors contributing to cognitive decline and dementia have been identified, sleep has only recently been highlighted as a potential modifiable risk factor. The chapter by Lam and colleagues discusses the association between sleep and circadian disruption with cognitive decline and dementia and potential neurobiological processes underpinning this link, as well as the implications for sleep intervention.

The impact of sleep in other health condition is discussed across Chapters 7–9, exploring the relationship between sleep, health and cognition. Chapter 7 by Mason and Saletin focuses on the impact of insufficient sleep and circadian disruption in adolescence on health and cognition (particularly academic performance), with important discussion of societal factors that may help to buffer these impacts. This is followed by a chapter by Rasmussen and colleagues outlining the interplay of sleep health and cardiometabolic disease. Finally, Bilterys, Nijs and Tang discuss the well-known bidirectional association between sleep and chronic pain, and its implications for the management of sleep problems in these patients.

The final set of chapters discusses the broader societal impacts of sleep and circadian disruption. It would be remiss to not acknowledge the enormous impact that our 24-7 society has placed on human health and well-being in the last century, with nearly 30% of the worldwide workforce now considered shift workers. Shen and colleagues discuss the impact of shift work on sleep and circadian rhythms and new digital approaches for managing the circadian response to shift work from an individual to an organisational level. Gaston and colleagues provide an important review of health disparities in sleep and mental health and discuss this from the perspective of traumatic childhood experiences and climate change.

The final chapter by Scott and Lechat provides an extensive review of the history of sleep monitoring, as well as current technologies and devices used to track and assess neurobiological underpinnings of sleep and circadian rhythms.

Together, these reviews highlight recent progress in understanding the behavioural neuroscience of sleep and identify promising areas for future research, including the possibility of sleep-based interventions to improve psychological and physical health.

These chapters are authored by experts in the field, who have a strong track record of research publications in the relevant subject area. We thank these wonderful colleagues for their invaluable contributions to this book. *We hope this book inspires you to explore these research and clinical avenues, and to consider sleep in all aspects of cognition, as well as mental and physical health.*

References

1 Patrick GTW, Gilbert JA. Studies from the psychological laboratory of the University of Iowa: On the effects of loss of sleep. *Psychological Review* 1896;3:469–83.
2 Aserinsky E, Kleitman N. Regularly occurring periods of eye motility, and concomitant phenomena, during sleep. *Science*. 1953;118:273–4.
3 Jackson ML, Van Dongen HPA. Cognitive effects of sleepiness. In: Thorpy M, Billiard M, editors. *Sleepiness*. Cambridge: Cambridge University Press; 2011. p. 72–81.

1

GLOBAL AND LOCAL SLEEP

Thomas Andrillon and Delphine Oudiette

Defining sleep

Behaviour, physiology and subjective experience

Sleep can be defined in many different ways, for example, using behavioural, metabolic, physiological, homeostatic, genetic signatures (1). In humans, behaviour, physiology and phenomenology are mainly used. At the behavioural level, sleep is operationally defined as transient episodes of unresponsiveness, and overt behaviour during sleep is usually thought to reflect arousal or abnormal sleep (e.g., parasomnia) (2).

At the physiological level, sleep is defined by clear changes in brain dynamics as well as muscular tone, ocular movements, heart rate, etc. (2, 3). There are two different types of sleep: non-rapid eye movement (NREM) and rapid eye movement (REM) sleep (2, 3). NREM sleep is characterised by slow waves, K-complexes or sleep spindles (4, 5). REM sleep is characterised by wake-like brain activity (EEG dominated by low-amplitude, high-frequency desynchronised rhythms) coupled with muscular atonia and phasic muscular events such as the eponymous rapid eye movements (6, 7).

Finally, sleep in humans can also be viewed through its associated phenomenological properties, especially when considering our first-person experience of our own sleep. Indeed, waking up from sleep, individuals often report not only vivid mental experiences with little or no association with the immediate environment (i.e., dreams) but also emerging from unconsciousness (8).

There is a large overlap between the physiological, behavioural and phenomenological definitions of sleep. Changes at the neuronal level indicative of sleep are closely associated with a loss of responsiveness and reports of either

DOI: 10.4324/9781003296966-1

dreams or unconsciousness. In particular, early investigations of REM sleep emphasised a tight relationship between REM sleep and dreaming on the one hand and NREM sleep and the loss of consciousness on the other hand (9, 10). Yet, these three dimensions of sleep do not perfectly overlap. For example, if dreams are very frequent during REM sleep, they are also common in NREM sleep (8, 11). Similarly, even if infrequent, behavioural responses during sleep are not completely abolished after sleep onset (12–14) or even during consolidated sleep (15, 16). We argue that these apparent contradictions could be resolved by reconsidering the vision of sleep as a global phenomenon.

Unihemispheric sleep and parasomnias

For some animals, the behavioural immobility associated with sleep is incompatible with their ecological niche. Dolphins, for example, need to swim to reach the surface and breathe (17). Interestingly, dolphins show a unique form of sleep: unihemispheric sleep, whereby one hemisphere shows EEG markers of sleep and the other, markers of wakefulness (18, 19). Unihemispheric sleep has also been observed in birds (20, 21).

Humans do not show an equivalent of unihemispheric sleep. Yet, sleep and wakeful activity can be simultaneously observed in different brain regions under specific circumstances, like in the case of parasomnias (22–24). For example, in disorders of arousal (25), patients show signs of partial awakenings from deep NREM sleep (26, 27), typically over the motor cortex and cerebellum, whereas frontal and parietal cortices still show NREM slow waves (28–30). This could explain many of the automatic behaviours, poor recall and altered consciousness characteristic of these states (27, 31, 32). Local modulations of sleep slow waves have also been observed in patients even when no episode or no sign of arousal is observed (33). These pathologies show that sleep is not always a global all-or-nothing phenomenon.

Local modulations of sleep within sleep

Slow waves as a building block of sleep

Slow waves are arguably one of the most fundamental aspects of NREM sleep. Slow waves entail a phenomenon of cortical bistability, whereby assemblies of neurons transition from silent (OFF) and active (ON) states in synchrony (34, 35). This bistable dynamic perturbs continuous deterministic processes and cortico-cortical dialogue (36–38), leading to a loss of responsiveness and of consciousness (39).

Neuronal bistability has been observed in cortical slices (40, 41), implying that slow waves can be generated by local cortical networks (42). Accordingly, slow waves represent the default mode of activity of cortical networks (43). In vivo, human intracranial recordings have shown that most of the intracranially recorded slow waves are local, whereas global waves (visible across all recording sites at once) are the exception rather than the rule (44) (Figure 1.1). It has been eventually proposed that there are two types of slow waves: global slow waves triggered by subcortical inputs and local slow waves generated by small cortical networks (45).

The reappraisal of slow waves as local events also highlights the frequent discrepancies between what is observed at the scalp level (which is often considered

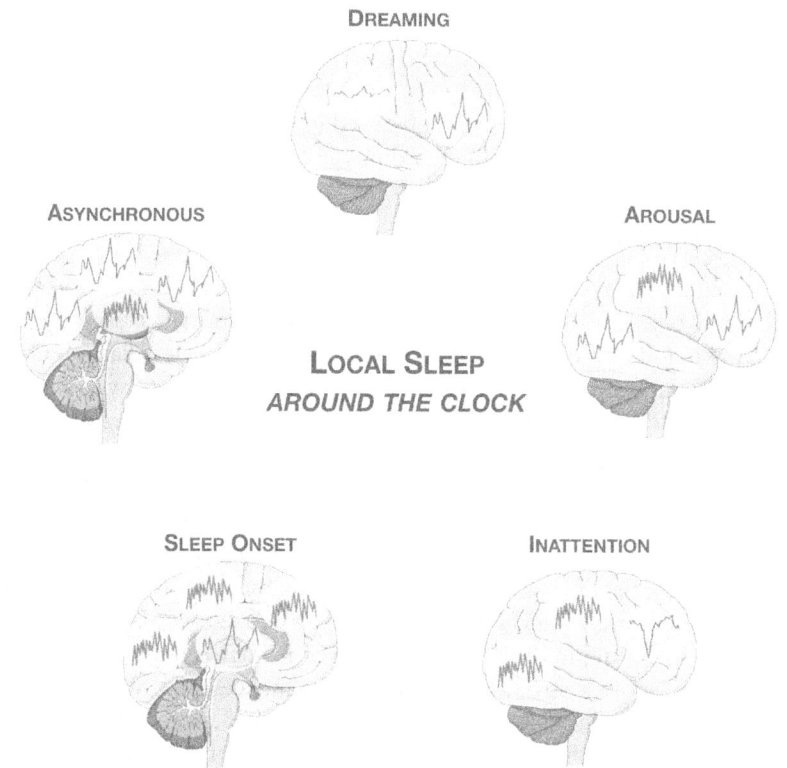

FIGURE 1.1   *Local aspects of wakefulness and sleep.* Different forms of local modulations of sleep have been evidenced. *Asynchronous* sleep refers to the coexistence of wake and sleep activity (as classically defined) in separate brain regions. At *sleep onset*, different brain regions transition to sleep at different times leading to windows of time, where some but not all of the brain has transitioned to sleep. During sleep itself, local modulations of slow-wave activity have been associated with *dreaming*, the feeling of being awake or the recovery of responsiveness. In the more contrasted case of (disorders of) *arousal*, an isolated brain region can show wake activity in the context of a globally sleeping brain. Conversely, during wake, slow waves can intrude in wakefulness, leading to moments of *inattention* with behavioural and phenomenological consequences.

the ground truth to determine the state of sleep or wakefulness) and the regional level. Indeed, some brain regions (e.g., motor areas) can recover a wake-like pattern of activity during sleep, while sleep-like activity is preserved at the scalp level (46–48). Likewise, when considering brain regions independently (e.g., hippocampus and neocortex), significant discrepancies have been reported (49, 50), with the neocortex spending a greater proportion of time in wakefulness or REM sleep compared to the hippocampus (50). Importantly, these episodes of asynchronous sleep, during which two brain regions display patterns of activity from different stages, can last up to ~30 minutes.

Differences between cortical and subcortical regions have also been reported at sleep onset, with up to ten minutes between the first recorded brain region to fall asleep (thalamus) and the last (posterior superior temporal gyrus) (51). Moreover, sleep spindles have been observed in the hippocampus prior to the detection of sleep onset on scalp recordings (52).

The regulation of sleep

Sleep is regulated by a homeostatic (S) and circadian (C) process (53). The homeostatic regulation of sleep means that sleep likelihood increases with time spent awake and decreases with time spent asleep. The circadian regulation implies that, independently of the time spent awake or asleep, wakefulness is more frequent during the day, Stage 3 NREM sleep during the first part of the night and REM sleep during the second.

The homeostatic and circadian pressure for sleep, and especially NREM sleep, can be tracked using slow-wave activity (SWA, power in the delta band derived from the EEG). SWA decreases exponentially during the night, in accordance with the dissipation of the homeostatic pressure (54) and increases with time spent awake (55, 56). Importantly, SWA displays some local variations and, for example, builds up faster over frontal regions (57). Frontal regions, thus, show a stronger homeostatic influence (58), whereas central and parietal electrodes show a stronger circadian influence (59). Thus, the homeostatic and circadian regulation of sleep affects cortical activity at a global and local scale (60).

In addition to the global influence of time spent awake (homeostatic) or time of the day (circadian), sleep is also regulated in a use-dependent fashion. For example, a difficult motor task is associated with enhanced local levels of SWA over the motor cortices during sleep compared to a simpler version of the same task (61). Similarly, tactile stimulations on the right hand during the day lead to an increase in SWA over the left motor cortices during sleep (62). Conversely, the immobilisation of an arm leads to a decrease in SWA over contralateral motor cortices (63, 64).

Yet local modulations of sleep are not only use-dependent since both the context and sleep environment can also impact sleep locally. Sleeping in a new environment (e.g., a sleep lab) can lead to an asymmetry in the amount of SWA between the two brain hemispheres, with the left hemisphere showing lower levels of SWA within the default mode network (65). This lateralisation could have a functional significance as it was associated with increased sensitivity to external stimuli.

These results are in favour of a complex and hierarchical modulation of sleep, with global and local determinants for sleep occurrence and intensity (60, 66).

Local modulations of sleep and subjective experience

Sleep is often accompanied by dreams (67). An association between dreaming and REM sleep, on the one hand, and dreamless sleep and NREM sleep, on the other hand, has been very influential (9). This association was supported by the fact that REM awakenings are often associated with dream reports, and the brain activity typically associated with REM sleep is closer to wakefulness (a state typically associated with consciousness) than NREM sleep (68). However, this distinction fails to account for the frequent occurrence of dreaming in NREM sleep (8, 11, 67, 69). Indeed, the original REM/dreaming association could be salvaged by hypothesising the existence of covert REM processes during NREM sleep, associated with NREM dreams (70). This covert REM sleep could be akin to the asynchronous sleep discussed above (50). Accordingly, increasing REM sleep pressure through REM sleep deprivation reinforced the dreamlike quality of NREM sleep experiences (71). However, since dreaming can disappear after forebrain lesions that spare the brainstem generators of REM sleep (69), REM activity does not appear sufficient for the occurrence of dreaming. It also does not seem necessary as the pharmacological suppression of REM sleep does not impact dreaming (72). NREM and REM dreams could instead have a common generator related to local modulations of sleep as they share common neural correlates, which consist of a relative decrease in SWA and an increase in the power in higher frequencies over parieto-occipital regions (73).

These local aspects of sleep could also play a key role in the subjective feeling of being asleep (74). Indeed, discrepancies between the objective (through physiology) and subjective (through self-reports) assessments of sleep are widespread (75–77). These discrepancies are particularly prevalent in insomnia or hypersomnia (77, 78). A recent study using a serial awakening paradigm suggests that individuals do not experience Stage N3 as the deepest sleep (79), despite this stage being classically referred to as deep sleep partly due to higher arousal thresholds (see (80)). At the EEG level, increases in the power at higher frequencies and in the density and amplitude of sleep spindles, particularly over frontal electrodes, were the best predictors of sleepers' feeling of being awake while asleep (79), suggesting again that local modulations of sleep can also account from sleepers' subjective experience.

Local modulations of sleep and sensory processing

The thalamic gating hypothesis posits that sleepers are unresponsive because external inputs are blocked at the level of the thalamus preventing them to reach cortical regions (81, 82). However, a complete gating of sensory inputs by the thalamus is not compatible with recordings of primary auditory cortices, showing a reliable encoding of auditory stimuli during sleep (83–85). In fact, the analysis of covert auditory processing during sleep via the analyses of EEG responses to sounds (e.g., event-related potentials (86)) revealed the maintenance of relatively complex and flexible processes during NREM and REM sleep (82, 86). Thus, sleepers can detect semantic, arithmetic or probabilistic violations (87–89), select and prepare motor responses (90, 91) and learn (92, 93). Likewise, sleepers show different neural responses to their own name (94) or the familiarity and emotional tone of a voice (95, 96).

More recently, lucid dreamers were shown to be able to perform voluntary actions in their dreams (97, 98) and even dialogue with experimenters from within their dream using ocular or facial codes (15). This ability could even be preserved, to a limited extent, in healthy sleepers in N2 and REM sleep (16). These findings beg the question of the mechanisms allowing or preventing sleepers from processing and/or responding to sensory inputs during sleep (82).

Local modulations of sleep appear here particularly relevant (Figure 1.1). Indeed, inter-hemispheric modulations of SWA when sleeping in a new environment could make the brain more responsive to external inputs (65).

In addition to these spatial modulations, the ability to respond (covertly or overtly) to sounds during sleep could also depend on the temporal dynamics of auditory responses during sleep as shown in a study examining cover motor responses to words presented during NREM and REM sleep (91). In NREM sleep, sounds typically trigger sleep rhythms (slow waves, spindles), which is interpreted as a mechanism protecting sleep (99, 100). Following this initial protective response, a local decrease in the power of slow waves and spindles takes place, which overlaps in time and space with motor preparation (91). Accordingly, K-complexes, an NREM slow wave that often occurs in response to sensory stimuli (100), could create windows of wakefulness within sleep (101) and allow sleepers to orientate their attention to auditory inputs (102).

At longer time scales, fluctuations of responsiveness and indexes of rich cognitive states during sleep suggest the existence of windows of reactivity to external stimuli within sleep (16). These windows could be organised in time via infra-slow fluctuations (~0.02 Hz) of sleep hallmarks such as sleep spindles (103, 104), leading to a shift between states of low and high susceptibility to external stimuli (103).

Local sleep intrusions in wakefulness

Characterising sleep intrusions in awake animals

Local slow waves can be observed during what is classically considered wakefulness (because of the maintenance of wake activity in most brain regions or the preservation of normal wake behaviour). These local irruptions of sleep activity in wakefulness, in the form of high-amplitude slow waves and/or reduction of neuronal firing rate, have been first evidenced in the visual cortex of monkeys (105) and the somatosensory cortex of rats (106, 107). Importantly, these sleep-like patterns of brain activity are homeostatically regulated (their occurrence increases with time spent awake) and are associated with impaired responsiveness (107, 108).

Electric stimulations during wakefulness in somatosensory cortices also revealed sleep-like responses, with notably an increase in the amplitude and slope of responses with time spent awake, whereas a decrease was observed with time spent asleep (109). The interpretation of these results is that, as the homeostatic sleep pressure builds up, cortical networks will more and more react to external stimuli with a sleep-like, bistable response. These changes in cortical responses correlate with changes in SWA and are resorbed after a period of sleep (110, 111).

High-amplitude slow waves paired with off-periods (i.e., moments of relative neuronal silencing) have also been observed in awake but sleep-deprived animals performing a task. Importantly, sleep intrusions in the frontal but not parietal cortices were associated with behavioural errors, suggesting region-specific effects on cognitive processes (112). Consequently, the presence of these off-periods could provide a simple mechanistic explanation for the behavioural errors often observed following sleep deprivation. Indeed, sudden off-periods in wakefulness could perturb the neural computations performed by a given brain region. If these operations are relevant to the task at hand, then off-periods would be shortly followed by an error. Importantly, this interpretation implies that off-periods can be observed in all brain regions as time spent awake increases, but only off-periods occurring in brain regions involved in a task will be predictive of behaviour in that task (region-specific effects) (113, 114).

Although previous findings strongly suggest that slow waves observed in sleep and wakefulness share the same generative mechanisms, it remains unclear if episodes of neuronal silencing are associated with a phenomenon of hyperpolarisation of cortical neurons as in sleep, which would require intracellular recordings (115). In addition, even if slow waves detected in LFP signals were associated with a significant drop in neuronal firing rate (112), a recent study suggests that, even during NREM sleep, some delta waves detected at the scalp level are not associated with neuronal silences or do not show a homeostatic regulation (116). A better characterisation of slow waves across multiple recording scales is needed to understand when slow waves are associated (or not) with a phenomenon of neuronal silencing.

Sleep-like slow waves in sleep-deprived humans

Sleep-deprived humans also show a shift of neuronal dynamics towards sleep, even when they are still awake. At the scalp level, sleep deprivation is associated with an increase in SWA or individual sleep-like slow waves (117–119). Slow waves also increase in a use-dependent fashion with more slow waves in regions under high use (e.g., motor cortices after hours in a driving simulator) (117). Finally, the increase in sleep-like activity predicts the behavioural impairment associated with fatigue, again in a region-specific fashion (117–120). Accordingly, the occurrence of slow waves during wakefulness can account for a wide range of behavioural or cognitive outcomes: motor impairment (117, 120), failures of response inhibition (118) or failures of emotional regulation (119). It is, thus, likely that these sleep-like slow waves play a key role in the cognitive consequences of sleep deprivation.

In intracranial recordings, single-neuron activity revealed that neurons from the mediotemporal lobe show a form of neuronal lapse following sleep deprivation: (1) they respond with delayed and weaker bursts to visual stimuli, (2) these changes typically occur in trials with lapses of attention (121). In addition, these neuronal lapses were associated with more power in the slow frequency band, potentially due to a weaker desynchronisation of endogenous oscillations.

Sleep-like slow waves and lapses of attention

Sleep-like slow waves have also been observed without any sleep deprivation (122, 123) and increase with time spent on task, or correlate with subjective and objective (e.g., pupil size) signs of fatigue (123). At the behavioural level, these slow waves can also predict the occurrence of attentional lapses (123). Importantly, slow waves can predict two different types of lapses: sluggish and impulsive responses (123). Regional effects were reported with frontal slow waves being associated with impulsivity and posterior slow waves with sluggishness. This is particularly interesting since sluggishness and impulsivity are two major consequences of sleep deprivation and both participate in increasing behavioural variability (124, 125). Local sleep intrusions could be responsible for both tendencies since the cognitive consequences of sleep intrusions would depend on the cognitive operations conducted by a given brain region (113). Accordingly, sleep intrusions in frontal cortices would perturb executive functions, leading to impulsivity, and sleep intrusions in posterior cortices could perturb sensory integration, leading to sluggish responses.

Furthermore, sleep intrusions were also associated with changes in subjective experiences associated with fatigue (123, 126). These fluctuations of subjective experience include mind wandering (i.e., thinking about something else than the task at hand) and mind blanking (i.e., thinking about nothing or not remembering any ongoing thoughts) (113). Both mind wandering and mind blanking increase following cognitive effort or sleep deprivation (127–129). Interestingly, sleep-like slow waves could predict occurrences of mind wandering and mind blanking (123). This association between local sleep intrusions within wakefulness and mind wandering could have benefits and foster creative thinking (130). Indeed, staggering at the borderland between wakefulness and sleep is associated with creative insights (131), potentially because a reduction in executive control during this period helps individuals to experience a state of freewheeling, hyper-associative thoughts. Accordingly, individuals with narcolepsy, who experience more of these hybrid sleep/wake states, appear particularly creative compared to the general population (132).

Finally, sleep-like slow waves appear to share the same neuromodulatory pathways that govern sleep/wake transitions since an increase in dopamine and noradrenaline (which promote wakefulness (133, 134)) decreases sleep-like slow waves, whereas an increase in serotonin (which could facilitate sleep (135)) increases sleep-like slow waves (136). Importantly, the same neuromodulators are involved in the regulation of top-down attention (137, 138) and are targeted by treatments of attentional deficits (139, 140). In conclusion, sleep-like intrusions could represent one of the mechanistic links that tie the neuromodulation of attention and vigilance (134, 137).

Conclusion

Sleep is a multidimensional phenomenon that impacts the body and brain, cognition and consciousness. Sleep is not monolithic, it is composed of different sub-states, and these sub-states can overlap, both in time and space. In this chapter, we have reviewed the evidence for such a local view of sleep, in animals and humans, healthy and pathological sleep. The picture that emerges is of a multilevel regulation of sleep by different exogenous and endogenous variables: the time spent awake, the position within the circadian cycle, the previous wake or sleep activity, the physiological state of a brain network or neuron or even the environmental context. These regulations are themselves implemented at different levels, local and global, making sleep exquisitely complex.

Local modulations of sleep have important implications for the brain's functions. During sleep, these modulations seem to shape our ability to interact with our environment and determine sleep quality and stability as well as our dreams. During wakefulness, these modulations could also explain fluctuations in attention and cognitive performance. Hybrid states between wakefulness and sleep could facilitate mind wandering and foster creative insights.

Local modulations of sleep could finally be a key element to fulfilling some critical functions of sleep such as the consolidation of memories, the maintenance of neural metabolic and synaptic homeostasis or the clearance of metabolic waste. Accordingly, local sleep could represent an adaptive phenomenon that is directly triggered by internal (e.g., synaptic or metabolic) changes and environmental variables. To explore these tantalising questions, a multidisciplinary, multilevel approach to sleep is necessary.

References

1 Pennisi E. The simplest of slumbers. Science. 2021;374(6567):526–9.
2 Carskadon MA, Dement WC. Normal human sleep: an overview. In: Principles and Practice of Sleep Medicine. Elsevier; 2005. p. 13–23.
3 Iber C, Ancoli-Israel S, Chesson A, Quan S. The AASM Manual for the Scoring of Sleep and Associated Events: Rules, Terminology and Technical Specifications. Westchester, IL: American Academy of Sleep Medicine; 2007.
4 De Gennaro L, Ferrara M. Sleep spindles: an overview. Sleep Med Rev. 2003;7:423–40.
5 Léger D, Debellemaniere E, Rabat A, Bayon V, Benchenane K, Chennaoui M. Slow-wave sleep: from the cell to the clinic. Sleep Med Rev. 2018;41:113–32.
6 Hobson JA, Pace-Schott EF. The cognitive neuroscience of sleep: neuronal systems, consciousness and learning. Nat Rev Neurosci. 2002;3(9):679–93.
7 Nir Y, Tononi G. Dreaming and the brain: from phenomenology to neurophysiology. Trends Cogn Sci. 2010;14:88–100.
8 Foulkes WD. Dream reports from different stages of sleep. J Abnorm Soc Psychol. 1962;65:14–25.
9 Hobson JA. The Dreaming Brain. New York: Basic Books; 1988.
10 Nir Y, Massimini M, Boly M, Tononi G. Sleep and consciousness. In: Cavanna AE, Nani A, Blumenfeld H, Laureys S, éditeurs. Neuroimaging of Consciousness [Internet]. Berlin and Heidelberg: Springer; 2013. p. 133–82.
11 Siclari F, Larocque JJ, Postle BR, Tononi G. Assessing sleep consciousness within subjects using a serial awakening paradigm. Front Psychol. 2013;4:542.
12 Canales-Johnson A, Beerendonk L, Blain S, Kitaoka S, Ezquerro-Nassar A, Nuiten S, et al. Decreased alertness reconfigures cognitive control networks. J Neurosci. 2020;40(37):7142–54.
13 Jagannathan SR, Bareham CA, Bekinschtein TA. Decreasing alertness modulates perceptual decision-making. J Neurosci. 2022;42(3):454–73.
14 Strauss M, Sitt JD, Naccache L, Raimondo F. Predicting the loss of responsiveness when falling asleep in humans. NeuroImage. 2022;251:119003.
15 Konkoly KR, Appel K, Chabani E, Mangiaruga A, Gott J, Mallett R, et al. Real-time dialogue between experimenters and dreamers during REM sleep. Curr Biol. 2021;S0960982221000592.
16 Türker B, Musat EM, Chabani E, Fonteix-Galet A, Maranci JB, Wattiez N, et al. Behavioral and brain responses to verbal stimuli reveal transient periods of cognitive integration of external world in all sleep stages. Neuroscience. 2022. http://biorxiv.org/lookup/doi/10.1101/2022.05.04.490484
17 Siegel JM. Do all animals sleep? Trends Neurosci. 2008;31(4):208–13.
18 Mukhametov LM, Supin AY, Polyakova IG. Interhemispheric asymmetry of the electroencephalographic sleep patterns in dolphins. Brain Res. 1977;134(3):581–4.
19 Mascetti, Gian Gastone GG. Unihemispheric sleep and asymmetrical sleep: behavioral, neurophysiological, and functional perspectives. Nat Sci Sleep. 2016;8:221–38.

20 Rattenborg NC, Amlaner CJ, Lima SL. Unilateral eye closure and interhemispheric EEG asymmetry during sleep in the pigeon (Columba livia). Brain Behav Evol. 2001;58(6):323–32.

21 Rattenborg NC, Ungurean G. The evolution and diversification of sleep. Trends Ecol Evol. 2022;S0169534722002531.

22 Mahowald MW, Schenck CH. Insights from studying human sleep disorders. Nature. 2005;437(7063):1279–85.

23 Siclari F, Tononi G. Local aspects of sleep and wakefulness. Curr Opin Neurobiol. 2017;44:222–7.

24 Scarpelli S, Alfonsi V, Gorgoni M. Parasomnias and disruptive sleep-related disorders: insights from local sleep findings. J Clin Med. 2022;11(15):4435.

25 Sateia MJ. International classification of sleep disorders-third edition. Chest. 2014;146(5):1387–94.

26 Castelnovo A, Lopez R, Proserpio P, Nobili L, Dauvilliers Y. NREM sleep parasomnias as disorders of sleep-state dissociation. Nat Rev Neurol. 2018;14(8): 470–81.

27 Idir Y, Oudiette D, Arnulf I. Sleepwalking, sleep terrors, sexsomnia and other disorders of arousal: the old and the new. J Sleep Res. 2022;31(4). https://onlinelibrary.wiley.com/doi/10.1111/jsr.13596

28 Bassetti C, Vella S, Donati F, Wielepp P, Weder B. SPECT during sleepwalking. Lancet. 2000;356(9228):484–5.

29 Terzaghi M, Sartori I, Tassi L, Didato G, Rustioni V, LoRusso G, et al. Evidence of dissociated arousal states during NREM parasomnia from an intracerebral neurophysiological study. Sleep. 2009;32(3):409–12.

30 Terzaghi M, Sartori I, Tassi L, Rustioni V, Proserpio P, Lorusso G, et al. Dissociated local arousal states underlying essential clinical features of non-rapid eye movement arousal parasomnia: an intracerebral stereo-electroencephalographic study. J Sleep Res. 2012;21(5):502–6.

31 Oudiette D, Leu S, Pottier M, Buzare MA, Brion A, Arnulf I. Dreamlike mentations during sleepwalking and sleep terrors in adults. Sleep. 2009;32(12):1621–7.

32 Uguccioni G, Golmard JL, de Fontréaux AN, Leu-Semenescu S, Brion A, Arnulf I. Fight or flight? Dream content during sleepwalking/sleep terrors vs rapid eye movement sleep behavior disorder. Sleep Med. 2013;14(5):391–8.

33 Castelnovo A, Riedner BA, Smith RF, Tononi G, Boly M, Benca RM. Scalp and source power topography in sleepwalking and sleep terrors: a high-density EEG study. Sleep. 2016;39(10):1815–25.

34 Steriade M. Neuronal Substrates of Sleep and Epilepsy. 1st éd. Vol. 1. Cambridge: Cambridge University Press; 2003. 536 p.

35 Vyazovskiy VV, Harris KD. Sleep and the single neuron: the role of global slow oscillations in individual cell rest. Nat Rev Neurosci. 2013;14(6):443–51.

36 Massimini M, Ferrarelli F, Huber R, Esser SK, Singh H, Tononi G. Breakdown of cortical effective connectivity during sleep. Science. 2005;309(5744):2228–32.

37 Massimini M, Ferrarelli F, Sarasso S, Tononi G. Cortical mechanisms of loss of consciousness: insight from TMS/EEG studies. Arch Ital Biol. 2012;150(2–3): 44–55.

38 Pigorini A, Sarasso S, Proserpio P, Szymanski C, Arnulfo G, Casarotto S, et al. Bistability breaks-off deterministic responses to intracortical stimulation during non-REM sleep. NeuroImage. 2015;112:105–13.

39 Tononi G, Massimini M. Why does consciousness fade in early sleep? Ann N Y Acad Sci. 2008;1129:330–4.

40 Hinard V, Mikhail C, Pradervand S, Curie T, Houtkooper RH, Auwerx J, et al. Key electrophysiological, molecular, and metabolic signatures of sleep and wakefulness revealed in primary cortical cultures. J Neurosci. 2012;32(36):12506–17.

41 Jewett KA, Taishi P, Sengupta P, Roy S, Davis CJ, Krueger JM. Tumor necrosis factor enhances the sleep-like state and electrical stimulation induces a wake-like state in co-cultures of neurons and glia. Eur J Neurosci. 2015;42(4):2078–90.

42 Krueger JM, Rector DM, Roy S, Van Dongen HP, Belenky G, Panksepp J. Sleep as a fundamental property of neuronal assemblies. Nat Rev Neurosci. 2008;9:910–19.

43 Sanchez-Vives MV. Slow wave activity as the default mode of the cerebral cortex. Arch Ital Biol. 2015;23.

44 Nir Y, Staba R, Andrillon T, Vyazovskiy VV, Cirelli C, Fried I, et al. Regional slow waves and spindles in human sleep. Neuron. 2011;70:153–69.

45 Siclari F, Bernardi G, Riedner BA, LaRocque JJ, Benca RM, Tononi G. Two distinct synchronization processes in the transition to sleep: a high-density electroencephalographic study. Sleep. 2014; 37(10): 1621–1637.

46 Nobili L, Ferrara M, Moroni F, De Gennaro L, Russo GL, Campus C, et al. Dissociated wake-like and sleep-like electro-cortical activity during sleep. NeuroImage. 2011;58(2):612–19.

47 Nobili L, De Gennaro L, Proserpio P, Moroni F, Sarasso S, Pigorini A, et al. Local aspects of sleep. In: Progress in Brain Research [Internet]. Elsevier; 2012 p. 219–32.

48 Sarasso S, Pigorini A, Proserpio P, Gibbs SA, Massimini M, Nobili L. Fluid boundaries between wake and sleep: experimental evidence from stereo-EEG recordings. Arch Ital Biol. 2015;23.

49 Durán E, Oyanedel CN, Niethard N, Inostroza M, Born J. Sleep stage dynamics in neocortex and hippocampus. Sleep. 2018;41(6): 10.1093/sleep/zsy060.

50 Guthrie RS, Ciliberti D, Mankin EA, Poe GR. Recurrent hippocampo-neocortical sleep-state divergence in humans. Proc Natl Acad Sci. 2022;119(44):e2123427119.

51 Magnin M, Rey M, Bastuji H, Guillemant P, Mauguiere F, Garcia-Larrea L. Thalamic deactivation at sleep onset precedes that of the cerebral cortex in humans. Proc Natl Acad Sci. 2010;107(8):3829–33.

52 Sarasso S, Proserpio P, Pigorini A, Moroni F, Ferrara M, De Gennaro L, et al. Hippocampal sleep spindles preceding neocortical sleep onset in humans. NeuroImage. 2014;86:425–32.

53 Borbély AA. A two process model of sleep regulation. Hum Neurobiol. 1982;1(3):195–204.

54 Achermann P, Borbely AA. Mathematical models of sleep regulation. Front Biosci. 2003;8:s683–93.

55 Cajochen C, Foy R, Dijk DJ. Frontal predominance of a relative increase in sleep delta and theta EEG activity after sleep loss in humans. Sleep Res Online SRO. 1999;2(3):65–9.

56 Finelli LA, Baumann H, Borbély AA, Achermann P. Dual electroencephalogram markers of human sleep homeostasis: correlation between theta activity in waking and slow-wave activity in sleep. Neuroscience. 2000;101(3):523–9.

57 Ferrara M. Regional differences of the human sleep electroencephalogram in response to selective slow-wave sleep deprivation. Cereb Cortex. 2002;12(7):737–48.

58 Ferrara M, De Gennaro L. Going local: insights from EEG and stereo-EEG studies of the human sleep-wake cycle. Curr Top Med Chem. 2011;11(19):2423–37.

59 Lazar AS, Lazar ZI, Dijk DJ. Circadian regulation of slow waves in human sleep: topographical aspects. NeuroImage. 2015;116:123–34.

60 Krueger JM. Sleep and circadian rhythms: evolutionary entanglement and local regulation. Neurobiol Sleep Circadian Rhythms. 2020;9:100052.

61 Huber R, Ghilardi MF, Massimini M, Tononi G. Local sleep and learning. Nature. 2004;430:78–81.

62 Kattler H, Dijk DJ, Borbély AA. Effect of unilateral somatosensory stimulation prior to sleep on the sleep EEG in humans. J Sleep Res. 1994;3(3):159–64.

63 Huber R, Ghilardi MF, Massimini M, Ferrarelli F, Riedner BA, Peterson MJ, et al. Arm immobilization causes cortical plastic changes and locally decreases sleep slow wave activity. Nat Neurosci. 2006;9:1169–76.

64 Murphy M, Huber R, Esser SA, Riedner B, Massimini M, Ferrarelli F, et al. The cortical topography of local sleep. Curr Top Med Chem. 2011;11(19):2438–46.

65 Tamaki M, Bang JW, Watanabe T, Sasaki Y. Night watch in one brain hemisphere during sleep associated with the first-night effect in humans. Curr Biol. 2016;26(9):1190–4.

66 Szymusiak R. Hypothalamic versus neocortical control of sleep. Curr Opin Pulm Med. 2010;16(6):530–5.

67 Scarpelli S, Alfonsi V, Gorgoni M, De Gennaro L. What about dreams ? State of the art and open questions. J Sleep Res. 2022;31(4).

68 Hobson JA. REM sleep and dreaming: towards a theory of protoconsciousness. Nat Rev Neurosci. 2009;10:803–813.

69 Solms M. Dreaming and REM sleep are controlled by different brain mechanisms. Behav Brain Sci. 2000;23(6):843–50; discussion 904–1121.

70 Nielsen TA. A review of mentation in REM and NREM sleep: "covert" REM sleep as a possible reconciliation of two opposing models. Behav Brain Sci. 2000;23(6):851–66; discussion 904–1121.

71 Nielsen T, Stenstrom P, Takeuchi T, Saucier S, Lara-Carrasco J, Solomonova E, et al. Partial REM-sleep deprivation increases the dream-like quality of mentation from REM sleep and sleep onset. Sleep. 2005;28(9):1083–9.

72 Oudiette D, Dealberto MJ, Uguccioni G, Golmard JL, Merino-Andreu M, Tafti M, et al. Dreaming without REM sleep. Conscious Cogn. 2012;21(3):1129–40.

73 Siclari F, Baird B, Perogamvros L, Bernardi G, LaRocque JJ, Riedner B, et al. The neural correlates of dreaming. Nat Neurosci. 2017;20(6):872–8.

74 Andrillon T. Sleep: feeling awake while asleep. Curr Biol. 2021;31(24):R1578–80.

75 Bonnet MH, Moore SE. The threshold of sleep: perception of sleep as a function of time asleep and auditory threshold. Sleep. 1982;5(3):267–76.

76 Schinkelshoek MS, de Wit K, Bruggink V, Fronczek R, Lammers GJ. Daytime sleep state misperception in a tertiary sleep centre population. Sleep Med. 2020;69:78–84.

77 Valko PO, Hunziker S, Graf K, Werth E, Baumann CR. Sleep-wake misperception: a comprehensive analysis of a large sleep lab cohort. Sleep Med. 2021;88:96–103.

78 Edinger JD, Krystal AD. Subtyping primary insomnia: is sleep state misperception a distinct clinical entity? Sleep Med Rev. 2003;7(3):203–14.

79 Stephan AM, Lecci S, Cataldi J, Siclari F. Conscious experiences and high-density EEG patterns predicting subjective sleep depth. Curr Biol. 2021;31(24):5487–500.e3.

80 Bonnet MH, Johnson LC, Webb WB. The reliability of arousal threshold during sleep. Psychophysiology. 1978;15(5):412–16.

81 McCormick DA, Bal T. Sensory gating mechanisms of the thalamus. Curr Opin Neurobiol. 1994;4:550–6.

82 Andrillon T, Kouider S. The vigilant sleeper: neural mechanisms of sensory (de) coupling during sleep. Curr Opin Physiol. 2020;15:47–59.

83 Hennevin E, Huetz C, Edeline JM. Neural representations during sleep: from sensory processing to memory traces. Neurobiol Learn Mem. 2007;87:416–40.

84 Sela Y, Vyazovskiy VV, Cirelli C, Tononi G, Nir Y. Responses in rat core auditory cortex are preserved during sleep spindle oscillations. Sleep. 2016;39(5):1069–82.

85 Hayat H, Marmelshtein A, Krom AJ, Sela Y, Tankus A, Strauss I, et al. Reduced neural feedback signaling despite robust neuron and gamma auditory responses during human sleep. Nat Neurosci. 2022;25(7):935–43.

86 Bastuji H. Evoked potentials as a tool for the investigation of human sleep. Sleep Med Rev. 1999;3(1):23–45.

87 Ibanez A, Lopez V, Cornejo C. ERPs and contextual semantic discrimination: degrees of congruence in wakefulness and sleep. Brain Lang. 2006;98:264–75.

88 Strauss M, Sitt JD, King JR, Elbaz M, Azizi L, Buiatti M, et al. Disruption of hierarchical predictive coding during sleep. Proc Natl Acad Sci. 2015;112(11):E1353–62.

89 Strauss M, Dehaene S. Detection of arithmetic violations during sleep. Sleep: zsy232. 2019;42(3).

90 Kouider S, Andrillon T, Barbosa LS, Goupil L, Bekinschtein TA. Inducing task-relevant responses to speech in the sleeping brain. Curr Biol. 2014;24(18):2208–14.

91 Andrillon T, Poulsen AT, Hansen LK, Leger D, Kouider S. Neural markers of responsiveness to the environment in human sleep. J Neurosci. 2016;36(24):6583–96.

92 Arzi A, Shedlesky L, Ben-Shaul M, Nasser K, Oksenberg A, Hairston IS, et al. Humans can learn new information during sleep. Nat Neurosci. 2012;15(10):1460–5.

93 Andrillon T, Pressnitzer D, Léger D, Kouider S. Formation and suppression of acoustic memories during human sleep. Nat Commun 2017;8(1).

94 Perrin F, Garcia-Larrea L, Mauguiere F, Bastuji H. A differential brain response to the subject's own name persists during sleep. Clin Neurophysiol. 1999;110:2153–64.

95 Blume C, del Giudice R, Lechinger J, Wislowska M, Heib DPJ, Hoedlmoser K, et al. Preferential processing of emotionally and self-relevant stimuli persists in unconscious N2 sleep. Brain Lang. 2017;167:72–82.

96 Blume C, Del Giudice R, Wislowska M, Heib DPJ, Schabus M. Standing sentinel during human sleep: continued evaluation of environmental stimuli in the absence of consciousness. NeuroImage. 2018;178:638–48.

97 LaBerge S, Baird B, Zimbardo PG. Smooth tracking of visual targets distinguishes lucid REM sleep dreaming and waking perception from imagination. Nat Commun. 2018;9(1).

98 Oudiette D, Dodet P, Ledard N, Artru E, Rachidi I, Similowski T, et al. REM sleep respiratory behaviours match mental content in narcoleptic lucid dreamers. Sci Rep. 2018;8(1):2636.

99 Halasz P. K-complex, a reactive EEG graphoelement of NREM sleep: an old chap in a new garment. Sleep Med Rev. 2005;9:391–412.

100 Halász P. The K-complex as a special reactive sleep slow wave – a theoretical update. Sleep Med Rev. 2016;29:34–40.

101 Destexhe A, Hughes SW, Rudolph M, Crunelli V. Are corticothalamic 'up' states fragments of wakefulness? Trends Neurosci. 2007;30(7):334–42.

102 Legendre G, Andrillon T, Koroma M, Kouider S. Sleepers track informative speech in a multitalker environment. Nat Hum Behav. 2019;3(3):274–83.

103 Lecci S, Fernandez LMJ, Weber FD, Cardis R, Chatton JY, Born J, et al. Coordinated infraslow neural and cardiac oscillations mark fragility and offline periods in mammalian sleep. Sci Adv. 2017;3(2):e1602026.

104 Lázár ZI, Dijk DJ, Lázár AS. Infraslow oscillations in human sleep spindle activity. J Neurosci Methods. 2019;316:22–34.

105 Pigarev IN, Nothdurft HC, Kastner S. Evidence for asynchronous development of sleep in cortical areas. Neuroreport. 1997;8(11):2557–60.

106 Rector DM, Topchiy IA, Carter KM, Rojas MJ. Local functional state differences between rat cortical columns. Brain Res. 2005;1047(1):45–55.

107 Rector DM, Schei JL, Van Dongen HPA, Belenky G, Krueger JM. Physiological markers of local sleep. Eur J Neurosci. 2009;29(9):1771–8.

108 Krueger JM, Nguyen JT, Dykstra-Aiello CJ, Taishi P. Local sleep. Sleep Med Rev. 2019;43:14–21.

109 Vyazovskiy VV, Cirelli C, Pfister-Genskow M, Faraguna U, Tononi G. Molecular and electrophysiological evidence for net synaptic potentiation in wake and depression in sleep. Nat Neurosci. 2008;11(2):200–8.

110 Cajochen C, Brunner DP, Krauchi K, Graw P, Wirz-Justice A. Power density in theta/alpha frequencies of the waking EEG progressively increases during sustained wakefulness. Sleep. 1995;18(10):890–4.

111 Vyazovskiy VV, Riedner BA, Cirelli C, Tononi G. Sleep homeostasis and cortical synchronization: II. A local field potential study of sleep slow waves in the rat. Sleep. 2007;30(12):1631–42.

112 Vyazovskiy VV, Olcese U, Hanlon EC, Nir Y, Cirelli C, Tononi G. Local sleep in awake rats. Nature. 2011;472:443–7.

113 Andrillon T, Windt J, Silk T, Drummond SPA, Bellgrove MA, Tsuchiya N. Does the mind wander when the brain takes a break? Local sleep in wakefulness, attentional lapses and mind-wandering. Front Neurosci. 2019;13. Disponible sur: www.frontiersin.org/article/10.3389/fnins.2019.00949/full

114 D'Ambrosio S, Castelnovo A, Guglielmi O, Nobili L, Sarasso S, Garbarino S. Sleepiness as a local phenomenon. Front Neurosci. 2019;13. Disponible sur: www.frontiersin.org/article/10.3389/fnins.2019.01086/full

115 Steriade M, Nunez A, Amzica F. Intracellular analysis of relations between the slow (< 1 Hz) neocortical oscillation and other sleep rhythms of the electroencephalogram. J Neurosci. 1993;13:3266–83.

116 El-Kanbi K, de Lavilléon G, Bagur S, Lacroix M, Benchenane K. Distinction between slow waves and delta waves sheds light to sleep homeostasis and their association to hippocampal sharp waves ripples. Neuroscience. 2022. biorxiv.org/lookup/doi/10.1101/2022.12.27.522034

117 Hung CS, Sarasso S, Ferrarelli F, Riedner B, Ghilardi MF, Cirelli C, et al. Local experience-dependent changes in the wake EEG after prolonged wakefulness. Sleep. 2013;36(1):59–72.

118 Bernardi G, Siclari F, Yu X, Zennig C, Bellesi M, Ricciardi E, et al. Neural and behavioral correlates of extended training during sleep deprivation in humans: evidence for local, task-specific effects. J Neurosci. 2015;35(11):4487–500.

119 Avvenuti G, Bertelloni D, Lettieri G, Ricciardi E, Cecchetti L, Pietrini P, et al. Emotion regulation failures are preceded by local increases in sleep-like activity. J Cogn Neurosci. 2021;33(11):2342–56.

120 Ahlstrom C, Jansson S, Anund A. Local changes in the wake electroencephalogram precedes lane departures. J Sleep Res. 2017;26(6):816–19.

121 Nir Y, Andrillon T, Marmelshtein A, Suthana N, Cirelli C, Tononi G, et al. Selective neuronal lapses precede human cognitive lapses following sleep deprivation. Nat Med. 2017;23(12):1474–80.

122 Quercia A, Zappasodi F, Committeri G, Ferrara M. Local use-dependent sleep in wakefulness links performance errors to learning. Front Hum Neurosci [Internet]. 2018;12 journal.frontiersin.org/article/10.3389/fnhum.2018.00122/full

123 Andrillon T, Burns A, MacKay T, Windt J, Tsuchiya N. Predicting lapses of attention with sleep-like slow waves. Nat Commun. 2021;12:3657.

124 Drummond SPA, Paulus MP, Tapert SF. Effects of two nights sleep deprivation and two nights recovery sleep on response inhibition. J Sleep Res. 2006;15(3):261–5.

125 O'Connell RG, Dockree PM, Robertson IH, Bellgrove MA, Foxe JJ, Kelly SP. Uncovering the neural signature of lapsing attention: electrophysiological signals predict errors up to 20 s before they occur. J Neurosci. 2009;29(26):8604–11.

126 Wienke C, Bartsch MV, Vogelgesang L, Reichert C, Hinrichs H, Heinze HJ, et al. Mind-wandering is accompanied by both local sleep and enhanced processes of spatial attention allocation. Cereb Cortex Commun. 2021;2(1):tgab001.

127 Poh JH, Chong PLH, Chee MWL. Sleepless night, restless mind: effects of sleep deprivation on mind wandering. J Exp Psychol Gen. 2016;145(10):1312–18.

128 Zhang Y, Kumada T. Relationship between workload and mind-wandering in simulated driving. Xu J, éditeur. PLoS ONE. 2017;12(5):e0176962.

129 Jubera-Garcia E, Gevers W, Van Opstal F. Local build-up of sleep pressure could trigger mind wandering: evidence from sleep, circadian and mind wandering research. Biochem Pharmacol. 2021;191:114478.

130 Lacaux C. A doorway into possibility. Science. 2022;378(6620):611.

131 Lacaux C, Andrillon T, Bastoul C, Idir Y, Fonteix-Galet A, Arnulf I, et al. Sleep onset is a creative sweet spot. Sci Adv. 2021;7(50):eabj5866.

132 Lacaux C, Izabelle C, Santantonio G, De Villèle L, Frain J, Lubart T, et al. Increased creative thinking in narcolepsy. Brain J Neurol. 2019;142(7):1988–99.

133 Jones BE. From waking to sleeping: neuronal and chemical substrates. Trends Pharmacol Sci. 2005;26(11):578–86.

134 Sara SJ. The locus coeruleus and noradrenergic modulation of cognition. Nat Rev Neurosci. 2009;10:211–23.

135 Oikonomou G, Altermatt M, Zhang R wei, Coughlin GM, Montz C, Gradinaru V, et al. The serotonergic raphe promote sleep in zebrafish and mice. Neuron. 2019;103(4):686–701.e8.

136 Pinggal E, Dockree PM, O'Connell RG, Bellgrove MA, Andrillon T. Pharmaco-logical manipulations of physiological arousal and sleep-like slow waves modulate sustained attention. J Neurosci. 2022;JN-RM-0836-22.

137 Sara SJ, Bouret S. Orienting and reorienting: the locus coeruleus mediates cogni-tion through arousal. Neuron. 2012;76(1):130–41.

138 Thiele A, Bellgrove MA. Neuromodulation of attention. Neuron. 2018;97(4):769–85.

139 Spencer SV, Hawk LW, Richards JB, Shiels K, Pelham WE, Waxmonsky JG. Stimulant treatment reduces lapses in attention among children with ADHD: the effects of methylphenidate on intra-individual response time distributions. J Abnorm Child Psychol. 2009;37(6):805–16.

140 Hvolby A. Associations of sleep disturbance with ADHD: implications for treat-ment. ADHD Atten Deficit Hyperact Disord. 2015;7(1):1–18.

2

ADVANCES IN SLEEP-ASSOCIATED MEMORY CONSOLIDATION RESEARCH

Ryan Bottary, Dan Denis, Bryan S. Baxter and Tony J. Cunningham

Sleep is widely believed to actively facilitate memory consolidation (1–4). Specifically, retention of newly acquired memories is superior across delays that include sleep compared to equivalent amounts of active wake (4, 5). Further, sleep-associated memory consolidation is theorised to be linked to a combination of neural mechanisms during initial memory encoding, coupled brain oscillations during sleep, tuning of signal-to-noise ratios via synaptic downscaling and a persistence of brain connectivity that facilitates memory retrieval after sleep (3).

In this chapter, we review current evidence supporting the memory-promoting effects of sleep. In particular, we discuss a contemporary understanding of the mechanisms of sleep-associated memory consolidation gained through advances in experimental methodology, sleep-related interventions and computational approaches. We aim to provide a balanced account of the literature by also describing recent findings that highlight the boundary conditions for sleep-related memory consolidation. Finally, we aim to synthesise seemingly contradictory findings and discuss potential future research directions to clarify the conditions under which sleep may optimise memory consolidation.

What is sleep?

Sleep has traditionally been defined as a reversible state of relative behavioural inactivity that corresponds with progressive reduction in responses to, and perception of, the external environment (6). In contemporary sleep science, sleep tends to be measured physiologically using polysomnography (PSG), the gold-standard measurement of brain oscillations, eye movements and muscle activity using a montage of surface electrodes. Using PSG recordings, sleep

DOI: 10.4324/9781003296966-2

is differentiated from wake and classified, or staged, into rapid eye movement (REM) sleep or one of three non-REM stages (N1, N2 or N3) (7). N3 is often referred to as Slow Wave Sleep (SWS), so we will use these terms interchangeably throughout this chapter. Over the course of a night, humans typically cycle through these four sleep stages roughly every 90 minutes, with the early part of the night enriched in SWS and later enriched in REM sleep. N1 sleep constitutes only a small portion of time spent asleep, while N2 sleep commonly comprises more than 50% of total sleep time (TST) and is equally distributed across the night. An example hypnogram, a graph of how sleep stages are distributed across the night, is shown in Figure 2.19(a).

Historically, sleep stages have been determined by visual inspection of PSG traces. To accomplish this, trained sleep scorers focus on the frequency of predominant brain oscillations (i.e., faster brain oscillations present during wake, REM and lighter stages of NREM sleep and slower brain oscillations present during deeper stages of NREM sleep) and discrete oscillatory events, including sawtooth waves during REM sleep and sleep spindles during N2 and SWS. Saccadic eye movements and reduced muscle tone are additional markers that are used to score REM sleep. To further characterise brain activity during sleep, signal processing techniques are applied to spectrally decompose predominant brain oscillations into their component frequencies (see Figure 2.1(b) and (c)). We will discuss the significance of both sleep macroarchitecture (i.e., time spent in certain sleep stages) and microarchitecture (i.e., discrete sleep oscillations and spectral power) in relation to memory processing in a later section.

Memory stages and categories

The lifecycle of a memory begins at encoding, when sensory information is processed into short-term storage, and progresses through consolidation, when newly encoded memories are integrated into long-term storage (8). Once memories are consolidated, they can be accessed (i.e., retrieved) as needed, updated and reconsolidated (9).

Memories have traditionally been divided into two general categories: declarative and non-declarative (10). Declarative memories are further subdivided into episodic memories, or memories tied to specific experiences (e.g., attending one's college graduation), and semantic memories, or memories for general information (e.g., what state your college was in). Non-declarative memories include memories for learned skills (e.g., how to play a musical instrument) and learned associations (e.g., that good musical performances tend to be followed by applause from an audience).

Memory consolidation can be understood at two, non-mutually exclusive levels. Synaptic consolidation refers to rapid, experience-dependent changes in synaptic weights (11). For example, synaptic connections between neurons become

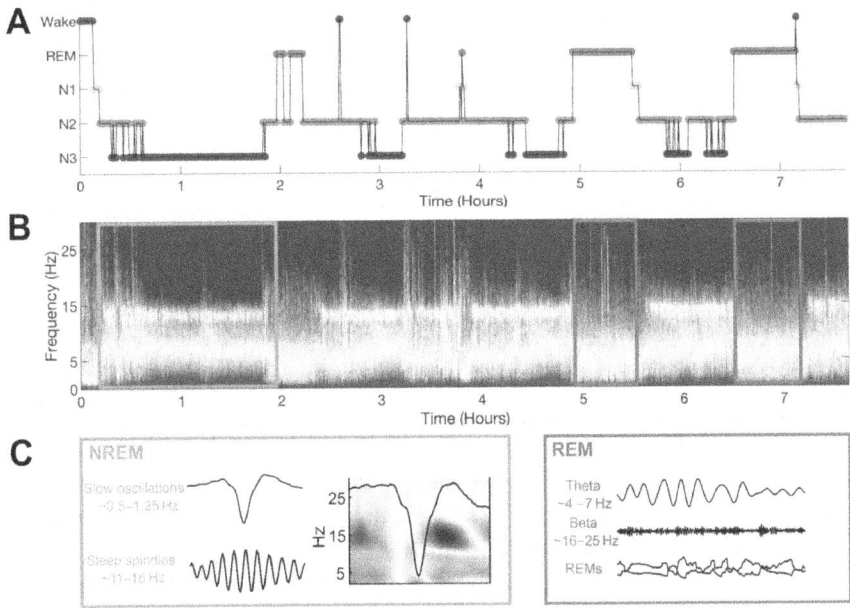

FIGURE 2.1 Overview of human sleep. (a) Hypnogram showing the cyclic occurrence of NREM sleep, comprising stages N1-N3 and REM sleep. Note the progressive decline of N3 sleep as the night progresses, and the relative increase in REM time later in the night. (b) Multitaper spectrogram of the sleep electroencephalogram (EEG) illustrating the oscillatory dynamics of the sleep EEG signal including a spectrum containing the canonical EEG frequencies (e.g. delta, 1–4 Hz; theta, 4–7 Hz; alpha, 8–12 Hz; sigma, 12–15 Hz; beta, 15–30 Hz; and gamma, >30 Hz). (c) Hallmark oscillatory events of sleep. Left: The most prominent oscillatory events during NREM sleep are neocortical slow oscillations and thalamocortical sleep spindles. The precise temporal coupling of sleep spindles [and hippocampal sharp-wave ripples, not shown here] to the excitable rising phase of the slow oscillation is believed to be a critical mechanism for sleep-related memory consolidation. Right: Oscillatory power in the theta (~4–7 Hz) and beta (~16–25 Hz) bands during REM sleep, as well as the number of rapid eye movements themselves, have all been proposed to serve a memory function, particularly in relation to highly salient emotional experiences.

stronger when those neurons are activated by new learning and this process, in turn, supports memory maintenance. Systems consolidation refers to changes in the distribution of a memory representation across brain regions (9, 12, 13). Specifically, systems consolidation refers to the process by which memories initially encoded and stored in the hippocampus are transferred to the neocortex for long-term storage. Beyond simply transferring carbon copies of newly encoded memories, hippocampal-neocortical dialogue is theorised to transform memory traces by abstracting relevant information and integrating this abstracted information into existing knowledge structures (9, 14). Importantly, systems consolidation is theorised to be a recursive process that may occur across both the short-term (several days and nights) and the long-term (several years) (15).

Sleep-associated memory consolidation

In the vast majority of studies, memories have been observed to be better remembered across delays including sleep compared to delays of active wakefulness (4, 5). This memory benefit by sleep was initially termed 'sleep-dependent memory consolidation' (16), though, as we will explore in later sections, memory consolidation does occur in the absence of sleep under certain circumstances. Therefore, we will instead use the term sleep-associated memory consolidation to reflect the memory benefit of sleep without discounting the fact that memory consolidation occurs during other states.

Sleep-associated memory consolidation is typically studied by having participants encode or learn novel information and then testing them on that information following a delay including some amount of sleep or wake. Those who sleep may do so overnight or during a daytime nap, while those who remain awake may do so across the day or overnight (i.e., sleep deprivation). Further, wake may be active (i.e., participants are engaged in ongoing daily tasks outside of the lab, in-lab tasks or deprived of sleep overnight) or non-active (e.g., quiet rest, or when participants remain awake, but do not engage in any tasks). Additional variants of sleep and memory studies include split-night studies (17), in which participants undergo learning, then sleep only during the first or second half of the night to take advantage of the natural changes in the distribution of SWS and REM, as well as selective deprivation of some sleep stages but not others (18, 19).

Importantly, sleep's memory-boosting benefit compared to active wake has been demonstrated across a full night of sleep and a daytime nap (20, 21). Further, sleep benefits declarative memory (e.g., picture memory, word-pair learning, visuospatial memory, memory for stories) by attenuating forgetting and non-declarative memory (e.g., fine and gross motor skills, associative learning) by improving performance compared to presleep baselines.

Theories supporting sleep-associated memory consolidation

Prioritised consolidation through memory tagging

How the brain selects which memories should be consolidated during sleep and what features of the memory lead to this selection, has been a topic of ongoing debate for over a decade (e.g., (22–26). Contemporary theories posit that newly encoded memories that have future relevance (i.e., are oriented toward goals such as avoiding threats or attaining positive outcomes) are tagged for selective consolidation during sleep. While the exact mechanisms of tagging are not yet fully characterised, tags may be set at the level of individual synapses and functionally connected brain networks (22, 24, 25).

Two types of memory tags have been proposed: encoding tags and retroactive tags (22). Encoding tags are set when newly encoded information has intrinsic salience (22, 24, 25). For example, emotional components of scenes (e.g., wrecked cars) tend to be remembered at the expense of their accompanying neutral backgrounds (e.g., buildings on the street where the car accident took place) (27, 28). Memory has also been shown to be superior for items associated with rewards and items that participants expect to later be tested on. However, whether sleep selectively consolidates tagged memories to a greater degree than wake remains uncertain (29). Retroactive tags are set when new information updates previously learned information, making that previously learned information now salient (22, 23). Evidence for retroactive tagging has come from studies demonstrating that pairing aversive (30) or rewarding (31) outcomes with new items within the same category as previously encoded items results in superior memory for original same-category items. Potential mechanisms supporting this selective consolidation of tagged memories during sleep will be discussed in the next sections.

Memory replay and selective consolidation during sleep

Sleep has been demonstrated to aid memory consolidation in two ways: replaying recently acquired memories in the hippocampus and strengthening long-term representations in the neocortex. Evidence for memory replay during sleep comes primarily from rodent studies demonstrating that neurons active during learning fire in the same sequence, but in a compressed time series, during hippocampal sharp-wave ripples during NREM sleep (3, 32). Preliminary evidence in humans has similarly demonstrated memory replay during sleep (33). Systems consolidation during sleep appears to be facilitated through coordinated NREM sleep oscillations (3). Specifically, successful memory consolidation has been linked to hippocampal sharp-wave ripples (SW-R; 150–200 Hz), thalamocortical sleep spindles (11–16 Hz) and cortical slow oscillations (SO, <1 Hz). Each of these oscillations has been independently associated with memory retention, though their coordination appears essential for predicting memory performance (3). Through repeated hippocampal replay and cortical activation, general features of the new memories may be abstracted, updating previous schemas, allowing for adaptive memory traces that will guide behaviours when faced with novel, yet similar, situations in the future (22). While no direct evidence supports the selective replay of tagged memories in humans, this remains a prominent theoretical mechanism for how the brain prioritises some memories over others for subsequent systems consolidation (22, 24, 25).

Sleep may aid in the downscaling of unimportant memories

In addition to selectively strengthening certain memories, sleep may also be critical for ridding the brain of information deemed unimportant. One theory supporting this idea relates to synaptic homeostasis, the observation that synaptic connections in the brain exhibit a net increase during wakefulness and subsequent downscaling during sleep (34, 35). During ongoing encoding across a typical day, new synaptic connections are formed that contain a combination of signal (i.e., important information to be moved to long-term storage) and noise (i.e., unimportant information). During sleep, it is theorised that synapses associated with unimportant memories are downscaled, while those associated with important memories may be prioritised for replay and consolidation (36). Recent modelling work supports this elegant view (37), yet direct evidence in humans or non-human animals is necessary to validate this theory.

Recent advances in memory-modulating sleep interventions

The theories above have been largely informed by behavioural (i.e., comparing memory performance across periods of sleep and wake) and sleep electrophysiological data (i.e., the EEG oscillations measured from the human brain and neuronal activity in rodents). While additional techniques discussed below, including sleep deprivation, facilitated memory reactivation and noninvasive brain stimulation, have been studied for a number of years, recent innovations in the delivery of these techniques have proven fruitful for refining our current understanding of the neurophysiological mechanisms involved in sleep-associated memory consolidation. This section describes both well-studied memory-boosting sleep intervention techniques and the state-of-the-art methodologies pushing the field forward.

Sleep deprivation

One way to test the potentially causal memory function of sleep in general, or specific sleep stages, is to measure memory performance in individuals deprived of it. Experimentally manipulating sleep can be done in a number of ways. Total sleep deprivation (TSD) involves depriving participants of a sleep opportunity entirely, thus requiring them to remain awake for at least one night. Partial sleep restriction involves reducing the overnight sleep opportunity to fewer hours than the participant typically sleeps over a single night or several nights. The causal impact of specific sleep stages on memory has also been tested using sleep-stage specific deprivation (i.e., depriving participants of certain sleep stages, such as REM or N3 sleep, through forced awakenings) and split-night sleep (as described above). The majority of sleep deprivation studies to date have implemented TSD, split-night, or selective sleep stage deprivation, though increased attention is being given to partial sleep restriction designs as they may be more ecologically relevant to the typical sleep loss experienced in humans.

TSD both prior to and following encoding has been shown to impair both procedural and declarative memory (38). TSD, the night prior to encoding, may impact memory retention in a number of ways. First, TSD has been shown to impair attention and encoding processes, including altered functional brain activity in memory-supporting structures, such as the hippocampus, prefrontal cortex and parietal cortices during encoding (39, 40). As discussed, sleep is also hypothesised to downscale synaptic connections, freeing space for new memory-related connections to be formed (36). Thus, TSD prior to encoding may result in the persistence of a saturated brain state in which the potential for new synaptic connections is limited. TSD following memory encoding also impairs memory consolidation when comparing memory performance to individuals who were given an uninterrupted sleep opportunity (40). Beyond the absence of memory-promoting sleep processes, which will be discussed in more detail below, TSD following encoding may impair the persistence, or strengthening, of hippocampal-neocortical connections critical to memory maintenance and retrieval (41).

Selectively depriving participants of specific sleep stages has been shown to impact the consolidation of certain memory types. For example, selectively depriving participants of REM sleep results in impaired retention of memories for extinguished fear (42). Selectively depriving participants of SWS sleep, thus leaving N2 and REM sleep intact, resulted in better retention of emotional, but not neutral memories (43), though selective REM deprivation has also been shown to have little impact on emotional memory retention (44). Further, evidence from split-night sleep studies suggests that emotional memory retention is better across late-night sleep enriched with REM, compared to early night SWS-enriched sleep (45). With respect to SWS deprivation, findings have been mixed with some showing that selective SWS sleep deprivation impairs visuospatial memory recall (18), but not procedural or word-pair memory (19). In summary, future research can use these types of total, partial, or targeted sleep deprivation techniques systematically across the different phases of memory encoding and consolidation to better understand the strengths and limitations of the effect of sleep on memory processing and to potentially advance our understanding of the underlying mechanisms supporting this effect (46).

Targeted memory reactivation

As mentioned above, spontaneous reactivation of memories during sleep (i.e., memory replay) is thought to drive memory consolidation processes (47–50).

A powerful experimental technique, targeted memory reactivation (TMR), has allowed researchers to externally manipulate reactivation processes, allowing the selective strengthening of some memories over others (51, 52). In a typical TMR experiment, studied stimuli are paired with either a sound or olfactory (i.e., scent) cue during encoding. Then, some, but not all, of these cues are represented to participants during sleep before a final memory test of all stimuli after awakening (53). Meta-analytic evidence supports a substantial TMR benefit, where items cued during sleep are remembered better than uncued items (54). TMR is especially effective when delivered during NREM sleep (stages N2 or SWS) and enhances both declarative and non-declarative memories (54).

Current research is now seeking to determine the mechanisms behind the TMR effect. TMR cues presented during NREM sleep have been shown to increase functional brain activity in the hippocampus (55) and evoke memory-promoting NREM sleep oscillations, including slow oscillations and sleep spindles (56, 57), suggesting TMR may evoke memory reactivation and processes important for memory consolidation. This is supported by studies showing that learning-related brain signatures are evoked or increased by TMR presentation during sleep (58, 59) and that memory content can be decoded from TMR-evoked electroencephalogram (EEG) patterns (57, 60). Although a powerful experimental tool, an important caveat to this approach is that artificial memory cueing may or may not share the same memory-boosting brain mechanisms as spontaneous reactivation.

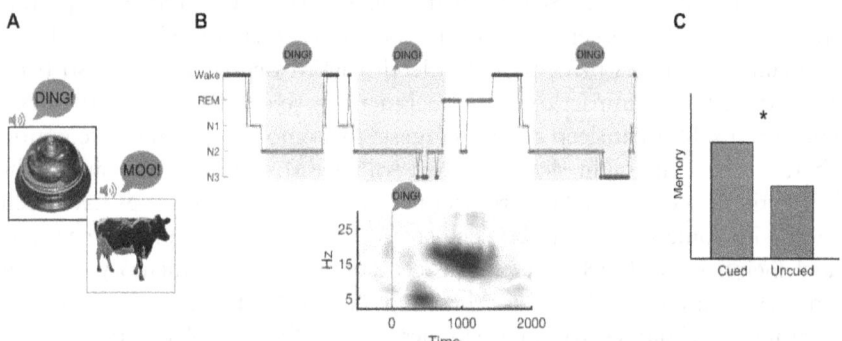

FIGURE 2.2   Example of a typical targeted memory reactivation (TMR) experiment. (a) During initial encoding, each to-be-learned stimulus is paired with a unique sound. (b) During N2 and/or N3 sleep, half of the sounds known as the cued items are replayed while the participant is asleep. These sounds evoke slow oscillation and spindle band activity (see lower panel). (c) Following sleep, memory for all the stimuli is tested. Typically, memory for the items that were cued during sleep is superior to memory for the items that were not played during sleep (the uncued condition).

Non-invasive brain stimulation

Non-invasive brain stimulation approaches have gained increasing interest in recent years as methods for boosting sleep signatures related to enhanced memory. Two techniques, auditory stimulation and transcranial electrical stimulation, will be discussed below. Importantly, in contrast to TMR, non-invasive brain stimulation is delivered during sleep only, without stimuli first being paired with studied items.

Auditory stimulation

Auditory stimulation is a technique that uses short, strategically timed noise bursts during sleep to evoke memory-promoting sleep oscillations without impairing the ability to record ongoing EEG activity. Closed-loop auditory stimulation, the most common approach implemented in studies aiming to improve memory, works by delivering short noise bursts during the up-state of slow oscillations in NREM sleep (see Figure 2.3). This approach has been

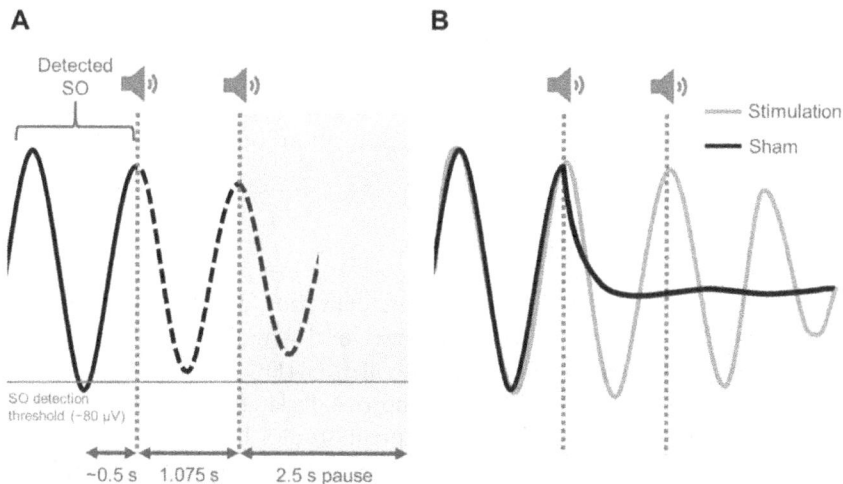

FIGURE 2.3 Example of a closed-loop stimulation protocol to enhance slow oscillatory activity (62). (a) In the stimulation condition, following detection of a slow oscillation downstate, two pulses of pink noise are delivered 50 ms apart and in phase with the two subsequent slow oscillation upstates. Following sound delivery, the detection routine is paused for 2.5 seconds. In the sham condition, corresponding time points of would-be stimulation are marked but no pulses are delivered. (b) Relative to sham, stimulation typically enhances the ongoing slow oscillation rhythm. Figure reproduced from Harrington and Cairney (76) under a Creative Commons Attribution 4.0 International License.

shown to improve declarative memory consolidation compared to sham stimulation in multiple studies (61–65), though see (66) for a counter-example. Auditory stimulation has been shown to increase slow oscillation power and fast-sigma power during the up-state of slow oscillations (61–64, 67) in both young and older adults, but stimulation may be less effective in older adults (68). Additionally, evoked slow oscillations (64, 69, 70) and fast-sigma power (62) positively correlate with memory performance following sleep. These electrophysiological increases have been shown to be robust to differences in methodology including composition of subjects (i.e., young adults and older adults) and type of sleep (i.e., nocturnal sleep versus daytime nap) (65).

Altering the timing or the frequency of the auditory stimulation alters electrophysiological and memory effects. For example, open-loop stimulation, delivering the stimuli randomly during NREM, or stimulating during the downstate of slow oscillations, does not increase sigma activity or improve memory consolidation (62, 71). Attempts have also been made to directly entrain spindles with sigma frequency auditory stimuli. Open-loop delivery of amplitude modulated white noise at sigma frequency increased spindles during stimulation, however, memory effects were not evaluated (72, 73). Closed-loop delivery of auditory stimuli at sigma frequency did not increase spindle density and did not improve declarative memory (74).

Only one study to date has attempted to promote REM theta activity using auditory stimulation (75). In this study, theta activity during REM sleep could be transiently increased using auditory stimulation, but this increase did not result in memory benefits.

Transcranial current stimulation

Transcranial current stimulation involves delivering a mild electric current to the scalp to either increase or decrease local neural activity. Transcranial direct current stimulation (tDCS) and oscillatory-tDCS at SO frequency during NREM sleep have been shown to improve declarative memory compared with sham stimulation (77–79). Subsequent studies have had mixed results with some finding improvements in declarative memory (80–82) and others no improvement relative to sham (83–86). Oscillatory-tDCS studies that also examined procedural memory were consistent in showing no effects. A meta-analysis including some of these studies determined an overall increase in declarative memory with stimulation and no effect on procedural memory (87). In contrast, in one study, transcranial alternating current stimulation (tACS), which is characterised by delivering the stimulation at a specific targeting frequency, at spindle frequency (12 Hz) time-locked to spindle detection improved performance on a motor procedural memory task, but not on a test of declarative memory (88). In another study, tACS time-locked to SOs and matched to their frequency and phase, increased SO power after stimulation,

SO-sigma phase-amplitude coupling and improved visual memory of a virtual environment (89). However, attempts to influence memory using theta frequency tDCS (5 Hz) have been unsuccessful (77). Similar to TMR, the effects generated by both auditory and electrical stimulation may or may not represent overlapping mechanisms of spontaneous reactivation with regard to brain mechanisms supporting memory consolidation. However, continued research using these techniques is valuable both in the potential of enhancing our understanding of the underlying mechanisms and as potential therapeutic tools to enhance markers of sleep and cognitive performance in both healthy and clinical cohorts.

Advances in the computational analysis of sleep and memory

A key tenet of the active systems consolidation hypothesis is that during sleep, neural patterns associated with learning are replayed in hippocampal and cortical sites (90). Although the measurement of replay of neuronal firing requires the use of invasive single unit recordings (33), advances in data analysis have allowed for broad patterns of memory reactivation to be detected non-invasively with scalp EEG and neuroimaging. In particular, machine learning classification algorithms allow for the decoding of the sleeping brain and have uncovered evidence of memory reprocessing during sleep. In one study, Schönauer and colleagues had participants encode either faces or houses before a night of sleep. A machine learning classifier was then trained to distinguish the two categories based on the sleep EEG patterns. They showed that it was possible to distinguish which category a participant had studied (faces or houses) based on the EEG recorded during sleep, suggesting that memory-related brain patterns emerge during sleep (48).

Using a similar approach (49), researchers asked whether the reactivation of memories uniquely occurred during slow oscillation-spindle complexes. Here, a classifier was trained to distinguish between objects and scenes studied while awake. Then, the classifier was applied to the sleep data. The researchers found significant above-chance reinstatement of learning patterns late in the SO-spindle complex, providing strong evidence that endogenous sleep oscillations promote memory reactivation and consolidation processes. Similar machine learning approaches have been used to identify reactivation of prior learning during slow wave sleep using fMRI (91), and hippocampal ripple events using intracranial EEG (92). New developments in electrode technology are increasing the resolution of invasive recording in clinical patients, which enables the simultaneous recording of hundreds of single units (93). The application of new analysis techniques, including deep learning, to electrophysiological and neuroimaging data will likely improve the detection of memory replay and allow for its evaluation in healthy and clinical cohorts, providing new opportunities for translational research.

Future directions

Several interesting studies in recent years have provided good evidence for sleep's active role in memory consolidation, though several outstanding questions remain. These include what sleep features drive memory benefits and whether the sleep state is both necessary and sufficient for memory consolidation.

Sleep stage and oscillatory correlates of memory consolidation are both routinely reported and highly controversial (94). Evidence from large-scale studies (95), meta-analyses (96) and individual empirical studies utilising commonly used behavioural tasks to probe sleep-associated memory consolidation (94) suggest that tracking memory consolidation to specific sleep stages or stage-specific oscillations is inconsistent at best. Therefore, future work, ideally in large samples using reliable behavioural tasks, is necessary to establish such associations. Advances in analytic techniques may also help us to move beyond current ways of conceptualising sleep states, providing insight into the dynamic interactions between sleep oscillations that support memory consolidation.

While the necessity of coupled hippocampal ripples, thalamocortical spindles and cortical slow oscillations for memory consolidation provides additional explanatory value, few studies in humans have directly measured such events and their relationship to the consolidation of different types of memories. Those that have are limited in their generalisability as the only human participants for which it is ethical to obtain hippocampal recordings from are those with neurological conditions such as epilepsy. Additional technological advancements are needed to break this barrier.

It will also be important to establish whether sleep is both *necessary* and *sufficient* to promote memory consolidation and systems consolidation specifically (97). For example, the idea that sleep initiates systems memory consolidation is challenged by work showing that neocortical memory engrams are formed from the outset of learning (98, 99) and that hippocampal independence of memory traces can be achieved in a single learning session (98). Further, intracranial recordings suggest that rather than the hippocampus initiating replay and dialogue with the neocortex through coordinated oscillations, prefrontal neocortical areas initiate which hippocampal traces are replayed (100). Although at least one study has demonstrated that the maintenance of rapid hippocampal independence during a single learning session requires post-learning sleep to be maintained across a 12-hour delay (101), whether hippocampal disengagement relates to sleep's impact on neocortical memory trace strengthening or hippocampal memory traces degradation remains unknown (97).

Emerging evidence also suggests that memory may be linked to specific neurophysiological correlates, irrespective of whether they occur in the sleep state or similar states of consciousness (e.g., quiet rest). Recent studies suggest that quiet rest, defined as task-free, unoccupied rest without detectable sleep (102), may result in memory benefits comparable to short periods of sleep (103–106). For example, Wang et al. (106) showed that memory performance for both declarative (i.e., word-pair learning) and procedural (i.e., sequential finger tapping) tasks was benefited by a short period of sleep or quiet rest, compared to a wake condition in which participants completed a distractor task. Importantly, the memory benefit was statistically equivalent for short sleep and quiet rest. As such, it may be that brain activity common to both quiet rest and sleep, such as memory replay, increased SO power, and hippocampal SW-Rs, are more important predictors of memory consolidation than the global sleep state per se (105).

Lastly, sleep has largely been observed or manipulated prior to or immediately following memory encoding. However, there are additional stages of memory processing that sleep may also be instrumental for, such as the extended consolidation period, prior to recall, and the reconsolidation of memories (46). For example, little work has been conducted to determine how sleep or sleep loss prior to retrieval of previously encoded memories impacts retrieval success. Additionally, sleep's role in memory updating (e.g., expanding memory schemas via reconsolidation) remains understudied. It would greatly benefit the field if future sleep and memory research takes an organised, stepwise approach to investigate the role of sleep at each phase of memory processing, thereby expanding our understanding of the role of sleep at all stages of the memory lifecycle. For a recent review, see (46).

Conclusions

Growing evidence supports an active role of sleep for the consolidation of a variety of memory types. Rather than promoting the storage of veridical memory traces at random, sleep may selectively strengthen and transform memories tagged during learning that serve an adaptive future purpose and discard memories with little future relevance. Several mechanisms for how sleep promotes memory consolidation have been proposed, including memory replay in the hippocampus and transfer to long-term storage in the neocortex via coupled hippocampal-neocortical sleep oscillations. Manipulating sleep, either by depriving participants of sleep or boosting brain processes and sleep oscillations, have been fruitful ways of both identifying causal links between sleep and memory consolidation, but also refining our understanding of the underlying neural mechanisms of this process. Despite these advances, recent work has begun to challenge whether sleep is both *necessary* and *sufficient* for memory consolidation. Future work exploring the strengths and limitations of the effect of sleep on the different stages and types of memory is necessary to address this question.

References

1 Born J, Wilhelm I. System consolidation of memory during sleep. Psychol Res. 2012;76(2):192–203.
2 Diekelmann S, Born J. The memory function of sleep. Nat Rev Neurosci. 2010;11(2):114–26.
3 Klinzing JG, Niethard N, Born J. Mechanisms of systems memory consolidation during sleep. Nat Neurosci. 2019;22(10):1598–610.
4 Rasch B, Born J. About sleep's role in memory. Physiol Rev. 2013;93(2):681–766.
5 Alger SE, Chambers AM, Cunningham T, Payne JD. The role of sleep in human declarative memory consolidation. In: Meerlo P, Benca RM, Abel T, editors. Sleep, Neuronal Plasticity and Brain Function. Berlin and Heidelberg: Springer; 2015. p. 269–306 (Current Topics in Behavioral Neurosciences). https://doi.org/10.1007/7854_2014_341
6 Carskadon MA. Normal human sleep: an overview In: Kryger M, Roth T, Dement WC, editors. Principles and Practice of Sleep Medicine, 6th ed. Philadelphia: Elsevier; 2017. p. 15–24.
7 Iber C, Ancoli-Israel S, Chesson AL, Quan SF. The AASM Manual for the Scoring of Sleep and Associated Events: Rules, Terminology and Technical Specifications. Vol. 1. Westchester, IL: American Academy of Sleep Medicine; 2007.
8 McGaugh JL. Memory – a century of consolidation. Science. 2000;287(5451): 248–51.
9 Moscovitch M, Gilboa A. Systems consolidation, transformation and reorganization: multiple trace theory, trace transformation theory and their competitors. PsyArXiv. 2021. Available from: https://psyarxiv.com/yxbrs/
10 Schacter DL, Tulving E. Memory Systems. Cambridge, MA: MIT Press; 1994.
11 Dudai Y, Karni A, Born J. The consolidation and transformation of memory. Neuron. 2015;88(1):20–32.
12 Craik FI. Remembering: an activity of mind and brain. Annu Rev Psychol. 2020;71:1–24.
13 Tonegawa S, Pignatelli M, Roy DS, Ryan TJ. Memory engram storage and retrieval. Curr Opin Neurobiol. 2015;35:101–9.
14 Moscovitch M, Cabeza R, Winocur G, Nadel L. Episodic memory and beyond: the hippocampus and neocortex in transformation. Annu Rev Psychol. 2016;67(1):105–34.
15 Dudai Y. The restless engram: consolidations never end. Annu Rev Neurosci. 2012;35(1):227–47.
16 Stickgold R. Sleep-dependent memory consolidation. Nature. 2005;437(7063): 1272–8.
17 Ackermann S, Rasch B. Differential effects of non-REM and REM sleep on memory consolidation? Curr Neurol Neurosci Rep. 2014;14(2):430.
18 Casey SJ, Solomons LC, Steier J, Kabra N, Burnside A, Pengo MF, et al. Slow wave and REM sleep deprivation effects on explicit and implicit memory during sleep. Neuropsychology. 2016;30(8):931–45.
19 Genzel L, Dresler M, Wehrle R, Grözinger M, Steiger A. Slow wave sleep and REM sleep awakenings do not affect sleep dependent memory consolidation. Sleep. 2009;32(3):9.
20 Berres S, Erdfelder E. The sleep benefit in episodic memory: an integrative review and a meta-analysis. Psychol Bull. 2021;147(12):1309.
21 Leong RL, Lo JC, Chee MW. Systematic review and meta-analyses on the effects of afternoon napping on cognition. Sleep Med Rev. 2022;101666.
22 Cowan ET, Schapiro AC, Dunsmoor JE, Murty VP. Memory consolidation as an adaptive process. Psychon Bull Rev. 2021. https://doi.org/10.3758/s13423-021-01978-x

23 Dunsmoor JE, Murty VP, Clewett D, Phelps EA, Davachi L. Tag and capture: how salient experiences target and rescue nearby events in memory. Trends Cogn Sci. 2022;26(9):782–95

24 Kim SY, Payne JD. Neural correlates of sleep, stress, and selective memory consolidation. Curr Opin Behav Sci. 2020;33:57–64.

25 Payne JD, Kensinger EA. Stress, sleep, and the selective consolidation of emotional memories. Curr Opin Behav Sci. 2018;19:36–43.

26 Stickgold R, Walker MP. Sleep-dependent memory triage: evolving generalization through selective processing. Nat Neurosci. 2013;16(2):139–45.

27 Cunningham TJ, Crowell CR, Alger SE, Kensinger EA, Villano MA, Mattingly SM, et al. Psychophysiological arousal at encoding leads to reduced reactivity but enhanced emotional memory following sleep. Neurobiol Learn Mem. 2014;114:155–64.

28 Kensinger EA, Garoff-Eaton RJ, Schacter DL. Effects of emotion on memory specificity: memory trade-offs elicited by negative visually arousing stimuli. J Mem Lang. 2007;56(4):575–91.

29 Davidson P, Jönsson P, Carlsson I, Pace-Schott E. Does sleep selectively strengthen certain memories over others based on emotion and perceived future relevance? Nat Sci Sleep. 2021;13:1257–306.

30 Dunsmoor JE, Murty VP, Davachi L, Phelps EA. Emotional learning selectively and retroactively strengthens memories for related events. Nature. 2015;520(7547):345–8.

31 Patil A, Murty VP, Dunsmoor JE, Phelps EA, Davachi L. Reward retroactively enhances memory consolidation for related items. Learn Mem. 2017;24(1):65–9.

32 Ji D, Wilson MA. Coordinated memory replay in the visual cortex and hippocampus during sleep. Nat Neurosci. 2007;10(1):100–7.

33 Rubin DB, Hosman T, Kelemen JN, Kapitonava A, Willett FR, Coughlin BF, et al. Learned motor patterns are replayed in human motor cortex during sleep. J Neurosci. 2022;42(25):5007–20.

34 Kuhn M, Wolf E, Maier JG, Mainberger F, Feige B, Schmid H, et al. Sleep recalibrates homeostatic and associative synaptic plasticity in the human cortex. Nat Commun. 2016;7(1):1–9.

35 Tononi G, Cirelli C. Sleep and the price of plasticity: from synaptic and cellular homeostasis to memory consolidation and integration. Neuron. 2014;81(1):12–34.

36 Cirelli C, Tononi G. Effects of sleep and waking on the synaptic ultrastructure. Philos Trans R Soc B Biol Sci. 2020;375(1799):20190235.

37 Robinson BS, Lau CW, New A, Nichols SM, Johnson EC, Wolmetz M, et al. Continual learning benefits from multiple sleep mechanisms: NREM, REM, and Synaptic Downscaling. arXiv. 2022. Available from: http://arxiv.org/abs/2209.05245

38 Newbury CR, Crowley R, Rastle K, Tamminen J. Sleep deprivation and memory: meta-analytic reviews of studies on sleep deprivation before and after learning. Psychol Bull. 2021;147(11):1215.

39 Drummond SPA, Brown GG, Gillin JC, Stricker JL, Wong EC, Buxton RB. Altered brain response to verbal learning following sleep deprivation. Nature. 2000;403(6770):655–7.

40 Krause AJ, Ben Simon E, Mander BA, Greer SM, Saletin JM, Goldstein-Piekarski AN, et al. The sleep-deprived human brain. Nat Rev Neurosci. 2017;18(7):404–18.

41 Chai Y, Fang Z, Yang FN, Xu S, Deng Y, Raine A, et al. Two nights of recovery sleep restores hippocampal connectivity but not episodic memory after total sleep deprivation. Sci Rep. 2020;10(1):8774.

42 Spoormaker VI, Schröter MS, Andrade KC, Dresler M, Kiem SA, Goya-Maldonado R, et al. Effects of rapid eye movement sleep deprivation on fear extinction recall and prediction error signaling. Hum Brain Mapp. 2011;33(10):2362–76.

43 Wiesner CD, Pulst J, Krause F, Elsner M, Baving L, Pedersen A, et al. The effect of selective REM-sleep deprivation on the consolidation and affective evaluation of emotional memories. Neurobiol Learn Mem. 2015;122:131–41.

44 Morgenthaler J, Wiesner CD, Hinze K, Abels LC, Prehn-Kristensen A, Göder R. Selective REM-sleep deprivation does not diminish emotional memory consolidation in young healthy subjects. PLoS ONE. 2014;9(2):e89849.

45 Schäfer SK, Wirth BE, Staginnus M, Becker N, Michael T, Sopp MR. Sleep's impact on emotional recognition memory: a meta-analysis of whole-night, nap, and REM sleep effects. Sleep Med Rev. 2020;51:101280.

46 Cunningham TJ, Stickgold R, Kensinger EA. Investigating the effects of sleep and sleep loss on the different stages of episodic emotional memory: a narrative review and guide to the future. Front Behav Neurosci. 2022;16:910317.

47 Pavlides C, Winson J. Influences of hippocampal place cell firing in the awake state on the activity of these cells during subsequent sleep episodes. J Neurosci. 1989;9(8):2907–18.

48 Schönauer M, Alizadeh S, Jamalabadi H, Abraham A, Pawlizki A, Gais S. Decoding material-specific memory reprocessing during sleep in humans. Nat Commun. 2017;8(1):15404.

49 Schreiner T, Petzka M, Staudigl T, Staresina BP. Endogenous memory reactivation during sleep in humans is clocked by slow oscillation-spindle complexes. Nat Commun. 2021;12(1):3112.

50 Wilson MA, McNaughton BL. Reactivation of hippocampal ensemble memories during sleep. Science. 1994;265(5172):676–9.

51 Lewis PA, Bendor D. How targeted memory reactivation promotes the selective strengthening of memories in sleep. Curr Biol. 2019;29(18):R906–12.

52 Oudiette D, Paller KA. Upgrading the sleeping brain with targeted memory reactivation. Trends Cogn Sci. 2013;17(3):142–9.

53 Rudoy JD, Voss JL, Westerberg CE, Paller KA. Strengthening individual memories by reactivating them during sleep. Science. 2009;326(5956):1079.

54 Hu X, Cheng LY, Chiu MH, Paller KA. Promoting memory consolidation during sleep: a meta-analysis of targeted memory reactivation. Psychol Bull. 2020;146(3):218.

55 Rasch B, Büchel C, Gais S, Born J. Odor cues during slow-wave sleep prompt declarative memory consolidation. Science. 2007;315(5817):1426–9.

56 Antony JW, Piloto L, Wang M, Pacheco P, Norman KA, Paller KA. Sleep spindle refractoriness segregates periods of memory reactivation. Curr Biol. 2018;28(11):1736–43.e4.

57 Cairney SA, Guttesen AÁV, El Marj N, Staresina BP. Memory consolidation is linked to spindle-mediated information processing during sleep. Curr Biol. 2018;28(6):948–54.e4.

58 Schreiner T, Doeller CF, Jensen O, Rasch B, Staudigl T. Theta phase-coordinated memory reactivation reoccurs in a slow-oscillatory rhythm during NREM sleep. Cell Rep. 2018;25(2):296–301.

59 Shanahan LK, Gjorgieva E, Paller KA, Kahnt T, Gottfried JA. Odor-evoked category reactivation in human ventromedial prefrontal cortex during sleep promotes memory consolidation. eLife. 2018;7:e39681.

60 Belal S, Cousins J, El-Deredy W, Parkes L, Schneider J, Tsujimura H, et al. Identification of memory reactivation during sleep by EEG classification. Neuroimage. 2018;176:203–14.

61 Leminen MM, Virkkala J, Saure E, Paajanen T, Zee PC, Santostasi G, et al. Enhanced memory consolidation via automatic sound stimulation during non-REM sleep. Sleep. 2017;40(3).

62 Ngo HVV, Martinetz T, Born J, Mölle M. Auditory closed-loop stimulation of the sleep slow oscillation enhances memory. Neuron. 2013;78(3):545–53.

63 Ong JL, Lo JC, Chee NIYN, Santostasi G, Paller KA, Zee PC, et al. Effects of phase-locked acoustic stimulation during a nap on EEG spectra and declarative memory consolidation. Sleep Med. 2016;20:88–97.

64 Papalambros NA, Santostasi G, Malkani RG, Braun R, Weintraub S, Paller KA, et al. Acoustic enhancement of sleep slow oscillations and concomitant memory improvement in older adults. Front Hum Neurosci. 2017;109.

65 Wunderlin M, Züst MA, Hertenstein E, Fehér KD, Schneider CL, Klöppel S, et al. Modulating overnight memory consolidation by acoustic stimulation during slow wave sleep – a systematic review and meta-analysis. Sleep. 2021;zsaa296.

66 Henin S, Borges H, Shankar A, Sarac C, Melloni L, Friedman D, et al. Closed-loop acoustic stimulation enhances sleep oscillations but not memory performance. eNeuro. 2019;6(6).

67 Lafon B, Henin S, Huang Y, Friedman D, Melloni L, Thesen T, et al. Low frequency transcranial electrical stimulation does not entrain sleep rhythms measured by human intracranial recordings. Nat Commun. 2017;8(1):1–14.

68 Navarrete M, Schneider J, Ngo HVV, Valderrama M, Casson AJ, Lewis PA. Examining the optimal timing for closed-loop auditory stimulation of slow-wave sleep in young and older adults. Sleep. 2020;43(6):zsz315.

69 Ong JL, Patanaik A, Chee NIYN, Lee XK, Poh JH, Chee MWL. Auditory stimulation of sleep slow oscillations modulates subsequent memory encoding through altered hippocampal function. Sleep. 2018;41(5).

70 Papalambros NA, Weintraub S, Chen T, Grimaldi D, Santostasi G, Paller KA, et al. Acoustic enhancement of sleep slow oscillations in mild cognitive impairment. Ann Clin Transl Neurol. 2019;6(7):1191–201.

71 Weigenand A, Mölle M, Werner F, Martinetz T, Marshall L. Timing matters: open-loop stimulation does not improve overnight consolidation of word pairs in humans. Eur J Neurosci. 2016;44(6):2357–68.

72 Antony JW, Paller KA. Using oscillating sounds to manipulate sleep spindles. Sleep. 2017;40(3).

73 Lustenberger C, Patel YA, Alagapan S, Page JM, Price B, Boyle MR, et al. High-density EEG characterization of brain responses to auditory rhythmic stimuli during wakefulness and NREM sleep. NeuroImage. 2018;169:57–68.

74 Ngo HVV, Seibold M, Boche DC, Mölle M, Born J. Insights on auditory closed-loop stimulation targeting sleep spindles in slow oscillation up-states. J Neurosci Methods. 2019;316:117–24.

75 Harrington MO, Ashton JE, Ngo HVV, Cairney SA. Phase-locked auditory stimulation of theta oscillations during rapid eye movement sleep. Sleep. 2021;44(4):zsaa227.

76 Harrington MO, Cairney SA. Sounding it out: auditory stimulation and overnight memory processing. Curr Sleep Med Rep. 2021;7(3):112–19.

77 Johnson JM, Durrant SJ. The effect of cathodal transcranial direct current stimulation during rapid eye-movement sleep on neutral and emotional memory. R Soc Open Sci. 2018;5(7):172353.

78 Marshall L. Transcranial direct current stimulation during sleep improves declarative memory. J Neurosci. 2004;24(44):9985–92.

79 Marshall L, Helgadóttir H, Mölle M, Born J. Boosting slow oscillations during sleep potentiates memory. Nature. 2006;444(7119):610–13.

80 Ladenbauer J, Külzow N, Passmann S, Antonenko D, Grittner U, Tamm S, et al. Brain stimulation during an afternoon nap boosts slow oscillatory activity and memory consolidation in older adults. Neuroimage. 2016;142:311–23.

81 Ladenbauer J, Ladenbauer J, Külzow N, de Boor R, Avramova E, Grittner U, et al. Promoting sleep oscillations and their functional coupling by transcranial stimulation enhances memory consolidation in mild cognitive impairment. J Neurosci. 2017;37(30):7111–24.

82 Westerberg CE, Florczak SM, Weintraub S, Mesulam MM, Marshall L, Zee PC, et al. Memory improvement via slow-oscillatory stimulation during sleep in older adults. Neurobiol Aging. 2015;36(9):2577–86.

83 Bueno-Lopez A, Eggert T, Dorn H, Danker-Hopfe H. Slow oscillatory transcranial direct current stimulation (so-tDCS) during slow wave sleep has no effects on declarative memory in healthy young subjects. Brain Stimul. 2019;12(4):948–58.

84 Eggert T, Dorn H, Sauter C, Nitsche MA, Bajbouj M, Danker-Hopfe H. No effects of slow oscillatory transcranial direct current stimulation (tDCS) on sleep-dependent memory consolidation in healthy elderly subjects. Brain Stimul. 2013;6(6):938–45.

85 Paßmann S, Külzow N, Ladenbauer J, Antonenko D, Grittner U, Tamm S, et al. Boosting slow oscillatory activity using tDCS during early nocturnal slow wave sleep does not improve memory consolidation in healthy older adults. Brain Stimul. 2016;9(5):730–9.

86 Sahlem GL, Badran BW, Halford JJ, Williams NR, Korte JE, Leslie K, et al. Oscillating square wave transcranial direct current stimulation (tDCS) delivered during slow wave sleep does not improve declarative memory more than sham: a randomized sham controlled crossover study. Brain Stimul. 2015;8(3):528–34.

87 Barham MP, Enticott PG, Conduit R, Lum JAG. Transcranial electrical stimulation during sleep enhances declarative (but not procedural) memory consolidation: evidence from a meta-analysis. Neurosci Biobehav Rev. 2016;63:65–77.

88 Lustenberger C, Boyle MR, Alagapan S, Mellin JM, Vaughn BV, Fröhlich F. Feedback-controlled transcranial alternating current stimulation reveals a functional role of sleep spindles in motor memory consolidation. Curr Biol. 2016;26(16):2127–36.

89 Ketz N, Jones AP, Bryant NB, Clark VP, Pilly PK. Closed-loop slow-wave tACS improves sleep-dependent long-term memory generalization by modulating endogenous oscillations. J Neurosci. 2018;38(33):7314–26.

90 Genzel L, Dragoi G, Frank L, Ganguly K, de la Prida L, Pfeiffer B, et al. A consensus statement: defining terms for reactivation analysis. Phil Trans R Soc B. 2020;375(1799):20200001.

91 Sterpenich V, van Schie MKM, Catsiyannis M, Ramyead A, Perrig S, Yang HD, et al. Reward biases spontaneous neural reactivation during sleep. Nat Commun. 2021;12(1):4162.

92 Zhang H, Fell J, Axmacher N. Electrophysiological mechanisms of human memory consolidation. Nat Commun. 2018;9(1):1–11.

93 Paulk AC, Kfir Y, Khanna AR, Mustroph ML, Trautmann EM, Soper DJ, et al. Large-scale neural recordings with single neuron resolution using Neuropixels probes in human cortex. Nat Neurosci. 2022;25(2):252–63.

94 Mantua J. Sleep physiology correlations and human memory consolidation: where do we go from here? Sleep. 2018;41(2):zsx204.

95 Ackermann S, Hartmann F, Papassotiropoulos A, de Quervain DJF, Rasch B. No associations between interindividual differences in sleep parameters and episodic memory consolidation. Sleep. 2015;38(6):951–9.

96 Cordi MJ, Rasch B. No evidence for intra-individual correlations between sleep-mediated declarative memory consolidation and slow-wave sleep. Sleep. 2021;44(8):zsab034.

97 Pöhlchen D, Schönauer M. Sleep-dependent memory consolidation in the light of rapid neocortical plasticity. Curr Opin Behav Sci. 2020;33:118–25.

98 Brodt S, Gais S, Beck J, Erb M, Scheffler K, Schönauer M. Fast track to the neocortex: a memory engram in the posterior parietal cortex. Science. 2018;362(6418):1045–8.

99 Kitamura T, Ogawa SK, Roy DS, Okuyama T, Morrissey MD, Smith LM, et al. Engrams and circuits crucial for systems consolidation of a memory. Science. 2017;356(6333):73–8.

100 Helfrich RF, Lendner JD, Mander BA, Guillen H, Paff M, Mnatsakanyan L, et al. Bidirectional prefrontal-hippocampal dynamics organize information transfer during sleep in humans. Nat Commun. 2019;10(1):3572.
101 Himmer L, Schönauer M, Heib DPJ, Schabus M, Gais S. Rehearsal initiates systems memory consolidation, sleep makes it last. Sci Adv. 2019;5(4):eaav1695.
102 Wamsley EJ. Memory consolidation during waking rest. Trends Cogn Sci. 2019;23(3):171–3.
103 Brokaw K, Tishler W, Manceor S, Hamilton K, Gaulden A, Parr E, et al. Resting state EEG correlates of memory consolidation. Neurobiol Learn Mem. 2016;130:17–25.
104 Humiston GB, Tucker MA, Summer T, Wamsley EJ. Resting states and memory consolidation: a preregistered replication and meta-analysis. Sci Rep. 2019;9(1):19345.
105 Wamsley EJ, Summer T. Spontaneous entry into an "offline" state during wakefulness: a mechanism of memory consolidation? J Cogn Neurosci. 2020;1–21.
106 Wang SY, Baker KC, Culbreth JL, Tracy O, Arora M, Liu T, et al. 'Sleep-dependent' memory consolidation? Brief periods of post-training rest and sleep provide an equivalent benefit for both declarative and procedural memory. Learn Mem. 2021;28(6):195–203.

3

THE NEUROBIOLOGY OF INSOMNIA

Constant factors and changes between day and night

Ellemarije Altena

Introduction

Insomnia is a psychiatric disorder and is currently diagnosed based on subjective complaints only: having difficulties with sleep onset, sleep maintenance or having early morning awakenings for more than three nights a week, while there is ample opportunity to sleep. Insomnia is only diagnosed when these complaints are accompanied by affected daytime functioning (social, professional), and is not better explained by, or only occurs in the context of, a different sleep disorder, mental disorder or substance abuse (American Psychiatric Association, 2013). Insomnia has been recognised as a risk factor for a wide range of health aspects such as cardiovascular problems and a lower life expectancy (Baglioni & Riemann, 2012; Burgos et al., 2006; Carroll et al., 2015; Spiegelhalder et al., 2011). Comorbid factors of insomnia such as hyperarousal, depression and stress can further contribute to affected sleep and affected daytime functioning, possibly as much as the long-term sleep disruption itself (Basta et al., 2007; Bonnet & Arand, 2010; Drake et al., 2014; Harvey, 2002; Jansson & Linton, 2007; Morin et al., 2003). In fact, meta-analyses and intervention studies have shown that insomnia treatment can prevent the future development of psychiatric problems such as depression and anxiety (Baglioni & Riemann, 2012; Cheng et al., 2021; Hertenstein et al., 2016).

Mechanisms of normal sleep

To understand the onset of insomnia and its effects on psychological and physiological aspects of sleep, we need to understand how normal sleep typically functions. In a healthy sleep-wake cycle, prolonged wakefulness leads to

DOI: 10.4324/9781003296966-3

increased sleep pressure. The build-up of sleep pressure is important since it optimises night-time sleep quantity and quality (Borbély, 1982). On a physiological level, several processes further enhance sleep quality and quantity. Right before sleep onset, levels of the stress hormone cortisol, as well as body temperature, typically drop, while levels of the sleep-inducing hormone melatonin increase. Before awakening, cortisol levels as well as body temperature increase and melatonin decreases. These sleep-inducing and wake-inducing mechanisms allow restful sleep and refreshed awakening. Disruptions of these cycles, for instance caused by long duration excessive stress, irregular sleep and wake patterns, alcohol, medication, heavy night-time meal intake or caffeine intake, can lead to insomnia (Saper et al., 2005).

Insomnia and hyperarousal

Throughout the 24-hour cycle, many patients with insomnia show signs of hyperarousal (Pérusse et al., 2013), which can be particularly apparent at sleep onset. Hyperarousal is a constantly increased level of arousal, independent of the presence of external stimuli (Bonnet & Arand, 1995; Perlis et al., 1997; Riemann et al., 2010). The onset of hyperarousal is thought to be typically marked by a stressful event, an accumulation of stressful events over time or chronic stressors, resulting in activation of the hypothalamic–pituitary–adrenal (HPA) axis (Dressle et al., 2022; Elder et al., 2023). Increased or dysregulated HPA axis activity can then negatively affect sleep in its own right, causing sleep fragmentation and sleep deprivation. When this negative cycle endures, individuals run the risk to develop chronic insomnia, with sleep problems remaining even if the initial stress from the event itself has already subsided (Buckley & Schatzberg, 2005; Dressle et al., 2022).

Although insomnia is thus related to alterations in HPA axis activity and stressful events are known to dysregulate the HPA axis, not much is known about HPA axis activity in acute insomnia (Elder et al., 2023). The hyperarousal theoretical model of insomnia, however, suggests that elevated cognitive, emotional and physiological activity (expressed as, for example, heightened autonomic or central nervous system activity) are important in the pathophysiology of both acute insomnia and chronic insomnia. Further research should investigate which hyperarousal features can be related to sleep aspects of insomnia alone or to comorbid disorders of insomnia, such as elevated levels of depression, anxiety and emotional reactivity (Bastien, 2020; Kalmbach et al., 2020; Vargas et al., 2020).

Can insomnia be detected in brain structure differences?

Insomnia has further been linked to differences in brain structure, although publications are scarce and often include a low sample size. Altena, Vrenken et al. (2010) found differences in the orbitofrontal and parietal grey matter density, with the orbitofrontal region being related to insomnia severity (Altena, Vrenken et al., 2010). Falgàs et al. (2021) found, in a larger group of participants, that higher insomnia scores were related to lower volumes of the right ventral orbitofrontal and temporo-parietal junction and the left insula (Falgàs et al., 2021). Bresser et al. investigated the connections between grey matter regions by focusing on white matter fractional anisotropy (FA) and found lower FA in the anterior internal capsule for insomnia. These FA values, particularly in the right hemisphere, were correlated with higher scores on the insomnia severity index (Bresser et al., 2020). Jespersen et al. investigated a sub-network of brain regions that included mainly fronto-subcortical connections, with the insula as a key region, and found reduced structural connectivity within this network. These last findings could, given the functionality of the brain regions and their connecting networks, point at underlying differences in interoception, emotional processing, stress responses and possibly play a role in alterations in the generation of slow-wave sleep (Jespersen et al., 2020). Additionally, insomnia is linked to cerebral asymmetry, which correlates to depressive and anxious symptoms (St-Jean et al., 2012) as well as insomnia severity (Provencher et al., 2020).

Insomnia and genetic factors

Genetic factors have been investigated for insomnia in numerous studies, with heritability of insomnia estimations ranging from 22% to 59% in adults, which remains stable over time (Barclay & Gregory, 2013; Gehrman et al., 2011, 2013; Lind & Gehrman, 2016). Particular gene expression processes can be linked to insomnia, while specific genetic factors were found in common between frequent insomnia symptoms and other health factors including restless legs syndrome, cardiometabolic, behavioural, psychiatric and reproductive traits. Evidence has shown a possible causal link between insomnia symptoms and coronary artery disease, depressive symptoms and subjective well-being (Lane et al., 2019). By applying a novel gene prioritisation strategy in almost 600,000 insomnia patients and comparing them to 1.5 million controls, Watanabe et al. identified insomnia genes that were particularly associated with metabolic and psychiatric pathways (Watanabe et al., 2022).

Insomnia: only a psychological condition?

One of the main mechanisms underlying successful insomnia treatment, typically entailing cognitive behavioural therapy for insomnia (CBT-I), is a change in dysfunctional beliefs and attitudes about sleep (Altena, 2022; Altena et al., 2023; Parsons et al., 2021; Schwartz & Carney, 2012). It, thus, seems logical to assume that the mere experience of insomnia can be considered as psychological only. In line with this, by changing the basic ideas about sleep and its consequences, the sleep problem can be treated. However, studies investigating insomnia applying neurophysiological measures, some of which were addressed above, have shown that insomnia can be related to changed patterns of brain activity before sleep, during sleep and during the daytime. In the following paragraphs, these neurobiological factors of insomnia will be divided in those observed before sleep onset, during sleep, at awakening and during wake.

The neurobiology of insomnia before sleep onset

Many insomnia patients experience difficulties falling asleep, being hampered by repetitive thoughts, rumination and also increased heart rate and higher muscle tension. Hyperarousal can be detected through physiological measures. Increased heart rate, altered galvanic skin response, temperature and heart rate variability differences have been the main findings when comparing those with and without insomnia before sleep onset (Bonnet & Arand, 2001; Lack et al., 2008; Lushington et al., 2000; Monroe, 1967; Perlis et al., 2017; Vgontzas et al., 1998). Further, through electro-encephalography (EEG), increased evoked response potentials (ERPs) (Bastien et al., 2008) and higher frequency EEG activity at the peri-onset of sleep have been observed in insomnia patients (Bastien et al., 2013; Fernandez-Mendoza et al., 2016; Turcotte et al., 2011) as well as increased EEG power (theta, gamma: (Cortoos et al., 2006; Perlis et al., 2001; Zhao et al., 2021). Those with higher levels of physiologically measurable hyperarousal are in fact at higher health risk in later life (Li et al., 2015).

Hyperarousal can enhance the continuation of sensory and information processing as well as long-term memory formation. These processes should slow down or be stopped to facilitate sleep onset as well as the wake-to-sleep transitioning (Bonnet & Arand, 2010). The physiological and cognitive expressions of hyperarousal can in fact enhance each other: the awakening effects of increased heart rate can enhance brain activity leading to continuous repetitive thoughts, interoception and rumination. As such, the insomnia patient can enter a vicious circle of physiological and cognitive hyperarousal enhancing each other. It is therefore important to break through this vicious circle as part of insomnia treatment (Riemann et al., 2010).

The neurobiology of insomnia during sleep

When insomnia patients do fall asleep, sleep quality can be affected, which can be observed in sleep architecture changes during particular sleep stages. In normal sleep, EEG power and metabolism levels should be lower than in wake. In insomnia patients, increased EEG power has been found as well as higher metabolism levels in several cortical brain regions during sleep (Nofzinger, 2004; Van Someren, 2021). Increases in EEG power have been particularly found during both REM and NREM sleep, with increases in the theta, alpha and sigma waves during NREM sleep and in the alpha and sigma waves during REM sleep (Perlis et al., 2017). In insomnia, most sleep fragmentation occurs in REM sleep (Riemann et al., 2012). When woken from REM sleep, insomnia patients report more previous awakenings than controls, which may suggest their vigilance levels were higher during sleep (Feige et al., 2018).

Elevated levels of the stress hormone cortisol, which is typically low during normal sleep, have further been found during sleep in insomnia (Vargas et al., 2018). The inhibitory neurotransmitter GABA is higher during normal sleep and has been suggested to play a role in sleep maintenance. In insomnia, GABA has been found to be lower, suggesting its natural inhibitory effect has been reduced (Morgan et al., 2012; Winkelman et al., 2008).

The neurobiology of insomnia at morning awakening

With normal ageing, sleep schedules are naturally shifted. For instance, adolescents' natural sleep schedule is delayed, with natural sleep onset at midnight and awakening times at 9 a.m., while older adults typically have an advanced sleep schedule (Vitiello, 2006). The normal ageing process can in fact reflect both a weakening of the sleep-promoting process and a strengthening of the wake-promoting process (Putilov et al., 2013). Insomnia can be characterised by early morning awakenings, occurring more frequently in older adults. Insomnia treatment can thus include sleep education, by informing patients about these natural ageing shifts in sleep schedules, and thus reducing worries about lack of sleep, before moving on to treatment of the other symptoms.

Very few studies report results on physiological measures related to insomnia as measured at morning awakening. Those that do, however, show that insomnia patients display elevated levels of morning cortisol, increased cortisol awakening responses and higher 24-hour cortisol levels (Elder et al., 2023; Grimaldi et al., 2021; Xia et al., 2013; Zhang et al., 2014). Furthermore, as measured by EEG, higher beta and gamma activity has been observed in insomnia patients after final morning awakening. Specifically, insomnia patients showed higher beta power at frontal, and higher beta and gamma power at posterior derivations in the morning than healthy controls. These findings confirm insomnia-related alterations in cortisol, linked to elevated HPA axis activity as a feature of insomnia disorder (Riemann et al., 2015).

To my knowledge, no data are available on other neuroimaging studies using fMRI, NIRS or another technique, focusing specifically on brain activity patterns in insomnia at morning awakening, such as through resting state brain activity. Often, however, neuroimaging studies do not report the time of day the study was performed. It would be very interesting to focus on fluctuations in brain activity patterns specifically related to insomnia throughout the 24-hour cycle (Fafrowicz et al., 2019; Jiang et al., 2016).

The neurobiology of insomnia during wakefulness

Part of the clinical diagnostic DSM-5 criteria for insomnia is altered daytime functioning (American Psychiatric Association, 2013). Insomnia patients often complain about affected memory functioning and problems with multi-tasking linked to their sleep problems, so numerous studies have focused on verifying these subjective complaints by including standard cognitive tests of memory, working memory and other executive functions (Backhaus et al., 2006; Bastien et al., 2003; Brownlow et al., 2020; Fernandez-Mendoza et al., 2010; Fortier-Brochu et al., 2012; Fortier-Brochu & Morin, 2014; Fulda & Schulz, 2001). Effects are not consistent and subtle at best, with memory problems (Fortier-Brochu & Morin, 2014) and executive functioning (Altena, Ramautar et al., 2010; Ballesio et al., 2019; Ferreira & de Almondes, 2014) most consistently affected in meta-analyses, though several studies find no effects on these objective tests. Variability in insomnia diagnostic criteria applied and type of cognitive tasks chosen have been identified as factors explaining contradictory results.

However, most standard neuropsychological tasks are developed to detect changes in neurodegenerative diseases, while neuropsychiatric conditions such as mood disorders (depression, anxiety) or insomnia may not only produce more subtle cognitive deficits but are also typically characterised by fluctuations in symptoms throughout the day and night (Könen et al., 2015). Even if particular cognitive tasks would therefore be able to detect more subtle deficits, the moment of assessment may still have a large influence on the outcome, which can be the topic of future studies.

Some studies have investigated neurobiological factors that could underlie insomnia-related behaviour or performance differences. Brain prefrontal metabolism has been found to be lower during wake in insomnia patients compared to controls, suggestive of an imbalance of brain activity during both sleep and wake in insomnia (Nofzinger, 2004; Van Someren, 2021). fMRI findings during resting state point at insomnia-related hyperarousal and increased vigilance, in particular by changes in the insula network (Chen et al., 2014; Kay & Buysse, 2017). Increased functional connectivity in a hippocampal – prefrontal network related to rumination has been found to be related to insomnia severity (Leerssen et al., 2019). Also during resting state, less functional connectivity variability between the anterior salience network and the left executive-control network may point at less flexible interactions between these networks in insomnia (Wei et al., 2020).

When executing a cognitive task in functional magnetic resonance imaging (fMRI) or functional near-infrared spectroscopy (FNIRS), those brain regions typically implicated in the task have shown less brain activity in insomnia patients, particularly in tasks activating prefrontal brain regions, requiring cognitive flexibility, planning or working memory (Altena et al., 2008; Drummond et al., 2013; Gong et al., 2022; Stoffers et al., 2014). In line with insomnia patients showing attentional bias when presented with sleep-related emotional stimuli (Barclay & Ellis, 2013; Harris et al., 2015; Lundh et al., 1997), those brain regions involved in emotional reactivity such as the amygdala (Baglioni et al., 2014) and the brain reward network (Sanz-Arigita et al., 2021) show increased activity and connectivity in insomnia patients compared to controls. A different effect is observed in the cortisol response: when presented with stress tests during daytime, those at increased risk of insomnia, but without current insomnia, show blunted cortisol responses (Reffi et al., 2022), which may be linked to an already high overall level of cortisol.

These findings may, though based on very few studies, offer neural support for affected reactivity to emotional stimuli, in particular to sleep-related information, while the opposite pattern is observed when presented with cognitive tasks, showing hypoactivity in prefrontal brain regions. Some of these effects on brain activity partially normalise after effective cognitive behavioural therapy as shown in a cognitive flexibility task (Altena et al., 2008) but not on a planning and working memory task (Stoffers et al., 2014). Effects of CBT-I might thus be brain region-dependent, with stronger effects on the prefrontal cortex, although these findings remain to be replicated.

Summary

In summary, even though psychological factors play an important role in the onset and maintenance of insomnia, the condition is also characterised by several neurobiological factors. Insomnia has been associated with differences in brain structure in regions and networks associated with interoception, emotional processing and stress responses (e.g., orbitofrontal cortex, insula networks). Genotypes identified to be related to insomnia play an important role in related health factors including metabolic and psychiatric factors. Insomnia is a strong risk factor to develop depression and cardiovascular problems, the last in particular in insomnia with hyperarousal.

While most bodily functions decrease before normal sleep onset, such as heart rate and EEG power, insomnia has been found to be linked to increased EEG power, higher body temperature and affected heart rate variability before sleep onset, phenomena that frequently co-occur with rumination and repetitive thoughts. The combination of these factors may each play a role in delayed sleep onset. During sleep, REM fragmentation is characteristic of insomnia as well as higher cortisol levels, higher EEG power and lower levels of the inhibitory neurotransmitter GABA. Higher cortical metabolism levels further contribute to the image of sleep disturbance by a brain that remains too active when it should be at rest.

At awakening, higher cortisol levels and EEG power characterise those with insomnia. The few functional neuroimaging studies performed in insomnia patients during daytime that include tasks show different results for emotional and cognitive stimuli presented. Emotional stimuli can provoke enhanced brain activity in the amygdala and brain reward network, while cognitive stimuli result in hypoactivity in task-related brain regions in insomnia. Resting state findings from studies performed at daytime point at constant vigilance activity, particularly through enhanced insula network activity.

From these findings, the image emerges of a constantly physiologically activated state in insomnia. However, studies with larger sample sizes remain sparse. Future studies should particularly focus on fluctuations of these physiological markers of insomnia throughout day and night, which should include functional neuroimaging studies. A focus should further lie on investigating the treatment effects of neurobiological factors of insomnia, which are currently particularly sparse.

References

Altena, E. (2022). Psychophysiological Mechanisms of CBT-I. In C. Baglioni, C. A. Espie, D. Riemann, European Sleep Research Society, European Insomnia Network, & European Academy for Cognitive Behavioural Therapy for Insomnia (Eds.), *Cognitive-behavioural therapy for insomnia (CBT-I) across the life span* (1st ed., pp. 51–61). Wiley. https://doi.org/10.1002/9781119891192.ch4

Altena, E., Ellis, J., Camart, N., Guichard, K., & Bastien, C. (2023). Mechanisms of cognitive behavioural therapy for insomnia. *Journal of Sleep Research*, e13860. https://doi.org/10.1111/jsr.13860

Altena, E., Ramautar, J. R., Van Der Werf, Y. D., & Van Someren, E. J. W. (2010). Do sleep complaints contribute to age-related cognitive decline? In *Progress in brain research* (Vol. 185, pp. 181–205). Elsevier. https://doi.org/10.1016/B978-0-444-53702-7.00011-7

Altena, E., Van Der Werf, Y. D., Sanz Arigita, E. J., Voorn, T. A., Rombouts, S. A. R. B., Kuijer, J. P. A., & Van Someren, E. J. W. (2008). Prefrontal hypoactivation and recovery in insomnia. *Sleep*, *31*(9), 1271–6.

Altena, E., Vrenken, H., Van Der Werf, Y. D., van den Heuvel, O. A., & Van Someren, E. J. W. (2010). Reduced orbitofrontal and parietal gray matter in chronic insomnia: A voxel-based morphometric study. *Biological Psychiatry*, *67*(2), 182–5. https://doi.org/10.1016/j.biopsych.2009.08.003

American Psychiatric Association. (2013). *Diagnostic and statistical manual of mental disorders* (5th ed.). American Psychiatric Association.

Backhaus, J., Junghanns, K., Born, J., Hohaus, K., Faasch, F., & Hohagen, F. (2006). Impaired declarative memory consolidation during sleep in patients with primary insomnia: Influence of sleep architecture and nocturnal cortisol release. *Biological Psychiatry*, *60*(12), 1324–30. https://doi.org/10.1016/j.biopsych.2006.03.051

Baglioni, C., & Riemann, D. (2012). Is chronic insomnia a precursor to major depression? Epidemiological and biological findings. *Current Psychiatry Reports*, *14*(5), 511–18. https://doi.org/10.1007/s11920-012-0308-5

Baglioni, C., Spiegelhalder, K., Regen, W., Feige, B., Nissen, C., Lombardo, C., Violani, C., Hennig, J., & Riemann, D. (2014). Insomnia disorder is associated with increased amygdala reactivity to insomnia-related stimuli. *Sleep*, *37*(12), 1907–17. https://doi.org/10.5665/sleep.4240

Ballesio, A., Aquino, M. R. J. V., Kyle, S. D., Ferlazzo, F., & Lombardo, C. (2019). Executive functions in insomnia disorder: A systematic review and exploratory meta-analysis. *Frontiers in Psychology*, 10, 101. https://doi.org/10.3389/fpsyg.2019.00101

Barclay, N. L., & Ellis, J. G. (2013). Sleep-related attentional bias in poor versus good sleepers is independent of affective valence. *Journal of Sleep Research*, 22(4), 414–21. https://doi.org/10.1111/jsr.12035

Barclay, N. L., & Gregory, A. M. (2013). Quantitative genetic research on sleep: A review of normal sleep, sleep disturbances and associated emotional, behavioural, and health-related difficulties. *Sleep Medicine Reviews*, 17(1), 29–40. https://doi.org/10.1016/j.smrv.2012.01.008

Basta, M., Chrousos, G. P., Vela-Bueno, A., & Vgontzas, A. N. (2007). Chronic insomnia and the stress system. *Sleep Medicine Clinics*, 2(2), 279–91. https://doi.org/10.1016/j.jsmc.2007.04.002

Bastien, C. H. (2020). Does insomnia exist without hyperarousal? What else can there be? *Brain Sciences*, 10(4), 225. https://doi.org/10.3390/brainsci10040225

Bastien, C. H., LeBlanc, M., Daley, M., & Morin, C. M. (2003). Cognitive performance and sleep quality in the elderly suffering from chronic insomnia relationship between objective and subjective measures. *Journal of Psychosomatic Research*, 11.

Bastien, C. H., St-Jean, G., Morin, C. M., Turcotte, I., & Carrier, J. (2008). Chronic psychophysiological insomnia: Hyperarousal and/or inhibition deficits? An ERPs investigation. *Sleep*, 31(6), 12.

Bastien, C. H., Turcotte, I., St-Jean, G., Morin, C. M., & Carrier, J. (2013). Information processing varies between insomnia types: Measures of N1 and P2 during the night. *Behavioral Sleep Medicine*, 11(1):56–72.

Bonnet, M. H., & Arand, D. L. (2001). Impact of activity and arousal upon spectral EEG parameters. *Physiology & Behavior*, 74(3):291–8. https://doi.org/10.1016/S0031-9384(01)00581-9

Bonnet, M. H., & Arand, D. L. (2010). Hyperarousal and insomnia: State of the science. *Sleep Medicine Reviews*, 14(1), 9–15. https://doi.org/10.1016/j.smrv.2009.05.002

Borbély, A. A. (1982). A two process model of sleep regulation. *Human Neurobiology*, 1(3), 195–204.

Bresser, T., Foster-Dingley, J. C., Wassing, R., Leerssen, J., Ramautar, J. R., Stoffers, D., Lakbila-Kamal, O., van den Heuvel, M., & van Someren, E. J. W. (2020). Consistent altered internal capsule white matter microstructure in insomnia disorder. *Sleep*, 43(8), zsaa031. https://doi.org/10.1093/sleep/zsaa031

Brownlow, J. A., Miller, K. E., & Gehrman, P. R. (2020). Insomnia and cognitive performance. *Sleep Medicine Clinics*, 15(1), 71–6. https://doi.org/10.1016/j.jsmc.2019.10.002

Buckley, T. M., & Schatzberg, A. F. (2005). On the interactions of the hypothalamic-pituitary-adrenal (HPA) axis and sleep: Normal HPA axis activity and circadian rhythm, exemplary sleep disorders. *The Journal of Clinical Endocrinology & Metabolism*, 90(5), 3106–14. https://doi.org/10.1210/jc.2004-1056

Burgos, I., Richter, L., Klein, T., Fiebich, B., Feige, B., Lieb, K., Voderholzer, U., & Riemann, D. (2006). Increased nocturnal interleukin-6 excretion in patients with primary insomnia: A pilot study. *Brain, Behavior, and Immunity*, 20(3), 246–53. https://doi.org/10.1016/j.bbi.2005.06.007

Carroll, J. E., Seeman, T. E., Olmstead, R., Melendez, G., Sadakane, R., Bootzin, R., Nicassio, P., & Irwin, M. R. (2015). Improved sleep quality in older adults with insomnia reduces biomarkers of disease risk: Pilot results from a randomized controlled comparative efficacy trial. *Psychoneuroendocrinology*, *55*, 184–92. https://doi.org/10.1016/j.psyneuen.2015.02.010

Chen, M. C., Chang, C., Glover, G. H., & Gotlib, I. H. (2014). Increased insula coactivation with salience networks in insomnia. *Biological Psychology*, *97*, 1–8. https://doi.org/10.1016/j.biopsycho.2013.12.016

Cheng, P., Casement, M. D., Kalmbach, D. A., Castelan, A. C., & Drake, C. L. (2021). Digital cognitive behavioral therapy for insomnia promotes later health resilience during the coronavirus disease 19 (COVID-19) pandemic. *Sleep*, *44*(4), zsaa258. https://doi.org/10.1093/sleep/zsaa258

Cortoos, A., Verstraeten, E., & Cluydts, R. (2006). Neurophysiological aspects of primary insomnia: Implications for its treatment. *Sleep Medicine Reviews*, *10*(4), 255–66. https://doi.org/10.1016/j.smrv.2006.01.002

Drake, C. L., Pillai, V., & Roth, T. (2014). Stress and sleep reactivity: A prospective investigation of the stress-diathesis model of insomnia. *Sleep*, *37*(8), 1295–304. https://doi.org/10.5665/sleep.3916

Dressle, R. J., Feige, B., Spiegelhalder, K., Schmucker, C., Benz, F., Mey, N. C., & Riemann, D. (2022). HPA axis activity in patients with chronic insomnia: A systematic review and meta-analysis of case – control studies. *Sleep Medicine Reviews*, *62*, 101588. https://doi.org/10.1016/j.smrv.2022.101588

Drummond, S. P. A., Walker, M., Almklov, E., Campos, M., Anderson, D. E., & Straus, L. D. (2013). Neural correlates of working memory performance in primary insomnia. *Sleep*, *36*(9), 1307–16. https://doi.org/10.5665/sleep.2952

Elder, G. J., Altena, E., Palagini, L., & Ellis, J. G. (2023). Stress and the hypothalamic – pituitary – adrenal axis: How can the COVID-19 pandemic inform our understanding and treatment of acute insomnia? *Journal of Sleep Research*. https://doi.org/10.1111/jsr.13842

Fafrowicz, M., Bohaterewicz, B., Ceglarek, A., Cichocka, M., Lewandowska, K., Sikora-Wachowicz, B., Oginska, H., Beres, A., Olszewska, J., & Marek, T. (2019). Beyond the low frequency fluctuations: Morning and evening differences in human brain. *Frontiers in Human Neuroscience*, *13*, 288. https://doi.org/10.3389/fnhum.2019.00288

Falgàs, N., Illán-Gala, I., Allen, I. E., Mumford, P., Essanaa, Y. M., Le, M. M., You, M., Grinberg, L. T., Rosen, H. J., Neylan, T. C., Kramer, J. H., & Walsh, C. M. (2021). Specific cortical and subcortical grey matter regions are associated with insomnia severity. *PLoS ONE*, *16*(5), e0252076. https://doi.org/10.1371/journal.pone.0252076

Feige, B., Nanovska, S., Baglioni, C., Bier, B., Cabrera, L., Diemers, S., Quellmalz, M., Siegel, M., Xeni, I., Szentkiralyi, A., Doerr, J.-P., & Riemann, D. (2018). Insomnia – perchance a dream? Results from a NREM/REM sleep awakening study in good sleepers and patients with insomnia. *Sleep*, *41*(5). https://doi.org/10.1093/sleep/zsy032

Fernandez-Mendoza, J., Calhoun, S., Bixler, E. O., Pejovic, S., Karataraki, M., Liao, D., Vela-Bueno, A., Ramos-Platon, M. J., Sauder, K. A., & Vgontzas, A. N. (2010). Insomnia with objective short sleep duration is associated with deficits in neuropsychological performance: A general population study. *Sleep*, *33*(4), 459–65. https://doi.org/10.1093/sleep/33.4.459

Fernandez-Mendoza, J., Li, Y., Vgontzas, A. N., Fang, J., Gaines, J., Calhoun, S. L., Liao, D., & Bixler, E. O. (2016). Insomnia is associated with cortical hyperarousal as early as adolescence. *Sleep*, *39*(5), 1029–36. https://doi.org/10.5665/sleep.5746

Ferreira, O. D. L., & de Almondes, K. M. (2014). The executive functions in primary insomniacs: Literature review. *Perspectivas en Psicología*, *11*, 10.

Fortier-Brochu, É., Beaulieu-Bonneau, S., Ivers, H., & Morin, C. M. (2012). Insomnia and daytime cognitive performance: A meta-analysis. *Sleep Medicine Reviews*, *16*(1), 83–94. https://doi.org/10.1016/j.smrv.2011.03.008

Fortier-Brochu, É., & Morin, C. M. (2014). Cognitive impairment in individuals with insomnia: Clinical significance and correlates. *Sleep*, *37*(11), 1787–98. https://doi.org/10.5665/sleep.4172

Fulda, S., & Schulz, H. (2001). Cognitive dysfunction in sleep disorders. *Sleep Medicine Reviews*, *5*(6), 423–45. https://doi.org/10.1053/smrv.2001.0157

Gehrman, P. R., Byrne, E., Gillespie, N., & Martin, N. G. (2011). Genetics of insomnia. *Sleep Medicine Clinics*, *6*(2), 191–202. https://doi.org/10.1016/j.jsmc.2011.03.003

Gehrman, P. R., Pfeiffenberger, C., & Byrne, E. M. (2013). The role of genes in the insomnia phenotype. *Sleep Medicine Clinics*, *8*(3), 323–31. https://doi.org/10.1016/j.jsmc.2013.04.005

Gong, H., Sun, H., Ma, Y., Tan, Y., Cui, M., Luo, M., & Chen, Y. (2022). Prefrontal brain function in patients with chronic insomnia disorder: A pilot functional near-infrared spectroscopy study. *Frontiers in Neurology*, *13*, 985988. https://doi.org/10.3389/fneur.2022.985988

Grimaldi, D., Reid, K. J., Papalambros, N. A., Braun, R. I., Malkani, R. G., Abbott, S. M., Ong, J. C., & Zee, P. C. (2021). Autonomic dysregulation and sleep homeostasis in insomnia. *Sleep*, *44*(6), zsaa274. https://doi.org/10.1093/sleep/zsaa274

Harris, K., Spiegelhalder, K., Espie, C. A., MacMahon, K. M. A., Woods, H. C., & Kyle, S. D. (2015). Sleep-related attentional bias in insomnia: A state-of-the-science review. *Clinical Psychology Review*, *42*, 16–27. https://doi.org/10.1016/j.cpr.2015.08.001

Harvey, A. G. (2002). A cognitive model of insomnia. *Behaviour Research and Therapy*, *40*(8), 869–93. https://doi.org/10.1016/S0005-7967(01)00061-4

Hertenstein, E., Johann, A., Baglioni, C., Spiegelhalder, K., & Riemann, D. (2016). Treatment of insomnia – A preventive strategy for cardiovascular and mental disorders. *Mental Health & Prevention*, *4*(2), 96–103. https://doi.org/10.1016/j.mhp.2016.02.005

Jansson, M., & Linton, S. J. (2007). Psychological mechanisms in the maintenance of insomnia: Arousal, distress, and sleep-related beliefs. *Behaviour Research and Therapy*, *45*(3), 511–21. https://doi.org/10.1016/j.brat.2006.04.003

Jespersen, K. V., Stevner, A., Fernandes, H., Sørensen, S. D., Van Someren, E., Kringelbach, M., & Vuust, P. (2020). Reduced structural connectivity in insomnia disorder. *Journal of Sleep Research*, *29*(1). https://doi.org/10.1111/jsr.12901

Jiang, C., Yi, L., Su, S., Shi, C., Long, X., Xie, G., & Zhang, L. (2016). Diurnal variations in neural activity of healthy human brain decoded with resting-state blood oxygen level dependent fMRI. *Frontiers in Human Neuroscience*, *10*. https://doi.org/10.3389/fnhum.2016.00634

Kalmbach, D. A., Buysse, D. J., Cheng, P., Roth, T., Yang, A., & Drake, C. L. (2020). Nocturnal cognitive arousal is associated with objective sleep disturbance and indicators of physiologic hyperarousal in good sleepers and individuals with insomnia disorder. *Sleep Medicine*, *71*, 151–60. https://doi.org/10.1016/j.sleep.2019.11.1184

Kay, D., & Buysse, D. (2017). Hyperarousal and beyond: New insights to the pathophysiology of insomnia disorder through functional neuroimaging studies. *Brain Sciences*, *7*(12), 23. https://doi.org/10.3390/brainsci7030023

Könen, T., Dirk, J., & Schmiedek, F. (2015). Cognitive benefits of last night's sleep: Daily variations in children's sleep behavior are related to working memory fluctuations. *Journal of Child Psychology and Psychiatry*, *56*(2), 171–82. https://doi.org/10.1111/jcpp.12296

Lack, L. C., Gradisar, M., Van Someren, E. J. W., Wright, H. R., & Lushington, K. (2008). The relationship between insomnia and body temperatures. *Sleep Medicine Reviews*, *12*(4), 307–17. https://doi.org/10.1016/j.smrv.2008.02.003

Lane, J. M., Jones, S. E., Dashti, H. S., Wood, A. R., Aragam, K. G., van Hees, V. T., Strand, L. B., Winsvold, B. S., Wang, H., Bowden, J., Song, Y., Patel, K., Anderson, S. G., Beaumont, R. N., Bechtold, D. A., Cade, B. E., Haas, M., Kathiresan, S., Little, M. A., . . . & Saxena, R. (2019). Biological and clinical insights from genetics of insomnia symptoms. *Nature Genetics*, *51*(3), 387–93. https://doi.org/10.1038/s41588-019-0361-7

Leerssen, J., Wassing, R., Ramautar, J. R., Stoffers, D., Lakbila-Kamal, O., Perrier, J., Bruijel, J., Foster-Dingley, J. C., Aghajani, M., & van Someren, E. J. W. (2019). Increased hippocampal-prefrontal functional connectivity in insomnia. *Neurobiology of Learning and Memory*, *160*, 144–50. https://doi.org/10.1016/j.nlm.2018.02.006

Li, Y., Vgontzas, A. N., Fernandez-Mendoza, J., Bixler, E. O., Sun, Y., Zhou, J., Ren, R., Li, T., & Tang, X. (2015). Insomnia with physiological hyperarousal is associated with hypertension. *Hypertension*, *65*(3), 644–50.

Lind, M., & Gehrman, P. (2016). Genetic pathways to insomnia. *Brain Sciences*, *6*(4), 64. https://doi.org/10.3390/brainsci6040064

Lundh, L.-G., Froding, A., Gyllenhammar, L., Broman, J.-E., & Hetta, J. (1997). Cognitive bias and memory performance in patients with persistent insomnia. *Scandinavian Journal of Behaviour Therapy*, *26*(1), 27–35. https://doi.org/10.1080/16506079708412033

Lushington, K., Dawson, D., & Lack, L. (2000). Core body temperature is elevated during constant wakefulness in elderly poor sleepers. *Sleep*, *23*(4), 1–7. https://doi.org/10.1093/sleep/23.4.1d

Monroe, L. J. (1967). Psychological and physiological differences between good and poor sleepers. *Journal of Abnormal Psychology*, *72*(3), 255–64. https://doi.org/10.1037/h0024563

Morgan, P. T., Pace-Schott, E. F., Mason, G. F., Forselius, E., Fasula, M., Valentine, G. W., & Sanacora, G. (2012). Cortical GABA levels in primary insomnia. *Sleep*, *35*(6), 807–14. https://doi.org/10.5665/sleep.1880

Morin, C. M., Rodrigue, S., & Ivers, H. (2003). Role of stress, arousal, and coping skills in primary insomnia. *Psychosomatic Medicine*, *65*(2), 259–67. https://doi.org/10.1097/01.PSY.0000030391.09558.A3

Nofzinger, E. A. (2004). Functional neuroimaging evidence for hyperarousal in insomnia. *American Journal of Psychiatry*, *161*(11), 2126–8. https://doi.org/10.1176/appi.ajp.161.11.2126

Parsons, C. E., Zachariae, R., Landberger, C., & Young, K. S. (2021). How does cognitive behavioural therapy for insomnia work? A systematic review and meta-analysis of mediators of change. *Clinical Psychology Review*, *86*, 102027. https://doi.org/10.1016/j.cpr.2021.102027

Perlis, M. L., Ellis, J. G., Kloss, J. D., & Riemann, D. W. (2017). Etiology and pathophysiology of insomnia. In *Principles and practice of sleep medicine* (pp. 769–84.e4). Elsevier. https://doi.org/10.1016/B978-0-323-24288-2.00082-9

Perlis, M. L., Giles, D. E., Mendelson, W. B., Bootzin, R. R., & Wyatt, J. K. (1997). Psychophysiological insomnia: The behavioural model and a neuro-cognitive perspective. *Journal of Sleep Research*, *6*(3), 179–88. https://doi.org/10.1046/j.1365-2869.1997.00045.x

Perlis, M. L., Kehr, E. L., Smith, M. T., Andrews, P. J., Orff, H., & Giles, D. E. (2001). Temporal and stagewise distribution of high frequency EEG activity in patients with primary and secondary insomnia and in good sleeper controls. *Journal of Sleep Research*, *10*(2), 93–104. https://doi.org/10.1046/j.1365-2869.2001.00247.x

Perlis, M. L., Smith, M. T., & Pigeon, W. R. (2005). Etiology and Pathophysiology of Insomnia. In M. Kryger, T. Roth, W.C. Dement (Eds.), *Principles and Practice of Sleep Medicine* (6th ed., pp. 714–25). Elsevier.

Pérusse, A. D., Turcotte, I., St-Jean, G., Ellis, J., Hudon, C., & Bastien, C. H. (2013). Types of primary insomnia: Is hyperarousal also present during napping? *Journal of Clinical Sleep Medicine*, *9*(12), 1273–80. https://doi.org/10.5664/jcsm.3268

Provencher, T., Fecteau, S., & Bastien, C. (2020). Patterns of intrahemispheric EEG asymmetry in insomnia sufferers: An exploratory study. *Brain Sciences*, *10*(12), 1014. https://doi.org/10.3390/brainsci10121014

Putilov, A., Münch, M., & Cajochen, C. (2013). Principal component structuring of the non-REM sleep EEG spectrum in older adults yields age-related changes in the sleep and wake drives. *Current Aging Science*, *6*(3), 280–93. https://doi.org/10.2174/18746098060314010120341 2

Reffi, A. N., Cheng, P., Kalmbach, D. A., Jovanovic, T., Norrholm, S. D., Roth, T., & Drake, C. L. (2022). Is a blunted cortisol response to stress a premorbid risk for insomnia? *Psychoneuroendocrinology*, *144*, 105873. https://doi.org/10.1016/j.psyneuen.2022.105873

Riemann, D., Nissen, C., Palagini, L., Otte, A., Perlis, M. L., & Spiegelhalder, K. (2015). The neurobiology, investigation, and treatment of chronic insomnia. *The Lancet Neurology*, *14*(5), 547–58. https://doi.org/10.1016/S1474-4422(15)00021-6

Riemann, D., Spiegelhalder, K., Feige, B., Voderholzer, U., Berger, M., Perlis, M., & Nissen, C. (2010). The hyperarousal model of insomnia: A review of the concept and its evidence. *Sleep Medicine Reviews*, *14*(1), 19–31. https://doi.org/10.1016/j.smrv.2009.04.002

Riemann, D., Spiegelhalder, K., Nissen, C., Hirscher, V., Baglioni, C., & Feige, B. (2012). REM sleep instability – A new pathway for insomnia? *Pharmacopsychiatry*, s-0031–1299721. https://doi.org/10.1055/s-0031-1299721

Sanz-Arigita, E., Daviaux, Y., Joliot, M., Dilharreguy, B., Micoulaud-Franchi, J.-A., Bioulac, S., Taillard, J., Philip, P., & Altena, E. (2021). Brain reactivity to humorous films is affected by insomnia. *Sleep*, *44*(9), zsab081. https://doi.org/10.1093/sleep/zsab081

Saper, C. B., Cano, G., & Scammell, T. E. (2005). Homeostatic, circadian, and emotional regulation of sleep. *The Journal of Comparative Neurology*, *493*(1), 92–8. https://doi.org/10.1002/cne.20770

Schwartz, D. R., & Carney, C. E. (2012). Mediators of cognitive-behavioral therapy for insomnia: A review of randomized controlled trials and secondary analysis studies. *Clinical Psychology Review*, *32*(7), 664–75. https://doi.org/10.1016/j.cpr.2012.06.006

Spiegelhalder, K., Fuchs, L., Ladwig, J., Kyle, S. D., Nissen, C., Voderholzer, U., Feige, B., & Riemann, D. (2011). Heart rate and heart rate variability in subjectively reported insomnia: Heart rate variability in primary insomnia. *Journal of Sleep Research*, *20*(1pt2), 137–45. https://doi.org/10.1111/j.1365-2869.2010.00863.x

St-Jean, G., Turcotte, I., & Bastien, C. H. (2012). Cerebral asymmetry in insomnia sufferers. *Frontiers in Neurology*, *3*. https://doi.org/10.3389/fneur.2012.00047

Stoffers, D., Altena, E., van der Werf, Y. D., Sanz-Arigita, E. J., Voorn, T. A., Astill, R. G., Strijers, R. L. M., Waterman, D., & Van Someren, E. J. W. (2014). The caudate: A key node in the neuronal network imbalance of insomnia? *Brain*, *137*(2), 610–20. https://doi.org/10.1093/brain/awt329

Turcotte, I., St-Jean, G., & Bastien, C. H. (2011). Are individuals with paradoxical insomnia more hyperaroused than individuals with psychophysiological insomnia? Event-related potentials measures at the peri-onset of sleep. *International Journal of Psychophysiology*, *81*(3), 177–90. https://doi.org/10.1016/j.ijpsycho.2011.06.008

Van Someren, E. J. W. (2021). Brain mechanisms of insomnia: New perspectives on causes and consequences. *Physiological Reviews*, *101*(3), 995–1046. https://doi.org/10.1152/physrev.00046.2019

Vargas, I., Nguyen, A. M., Muench, A., Bastien, C. H., Ellis, J. G., & Perlis, M. L. (2020). Acute and chronic insomnia: What has time and/or hyperarousal got to do with it? *Brain Sciences*, *10*(2), 71. https://doi.org/10.3390/brainsci10020071

Vargas, I., Vgontzas, A. N., Abelson, J. L., Faghih, R. T., Morales, K. H., & Perlis, M. L. (2018). Altered ultradian cortisol rhythmicity as a potential neurobiologic substrate for chronic insomnia. *Sleep Medicine Reviews*, *41*, 234–43. https://doi.org/10.1016/j.smrv.2018.03.003

Vgontzas, A. N., Tsigos, C., Bixler, E. O., Stratakis, C. A., Zachman, K., Kales, A., Vela-Bueno, A., & Chrousos, G. P. (1998). Chronic insomnia and activity of the stress system: A preliminary study. *Journal of Psychosomatic Research*, *45*(1), 21–31.

Vitiello, M. V. (2006). Sleep in normal aging. *Sleep Medicine Clinics*, *1*(2), 171–6. https://doi.org/10.1016/j.jsmc.2006.04.007

Watanabe, K., Jansen, P. R., Savage, J. E., Nandakumar, P., Wang, X., 23andMe Research Team, Agee, M., Aslibekyan, S., Auton, A., Bell, R. K., Bryc, K., Clark, S. K., Elson, S. L., Fletez-Brant, K., Fontanillas, P., Furlotte, N. A., Gandhi, P. M., Heilbron, K., Hicks, B., . . . Posthuma, D. (2022). Genome-wide meta-analysis of insomnia prioritizes genes associated with metabolic and psychiatric pathways. *Nature Genetics*, *54*(8), 1125–32. https://doi.org/10.1038/s41588-022-01124-w

Wei, Y., Leerssen, J., Wassing, R., Stoffers, D., Perrier, J., & Van Someren, E. J. W. (2020). Reduced dynamic functional connectivity between salience and executive brain networks in insomnia disorder. *Journal of Sleep Research*, *29*(2). https://doi.org/10.1111/jsr.12953

Winkelman, J. W., Buxton, O. M., Jensen, J. E., Benson, K. L., O'Connor, S. P., Wang, W., & Renshaw, P. F. (2008). Reduced brain GABA in primary insomnia: Preliminary data from 4T proton magnetic resonance spectroscopy (1H-MRS). *Sleep*, *31*(11), 1499–506. https://doi.org/10.1093/sleep/31.11.1499

Xia, L., Chen, G.-H., Li, Z.-H., Jiang, S., & Shen, J. (2013). Alterations in hypothalamus-pituitary-adrenal/thyroid axes and gonadotropin-releasing hormone in the patients with primary insomnia: A clinical research. *PLoS ONE*, *8*(8), e71065. https://doi.org/10.1371/journal.pone.0071065

Zhang, J., Lam, S.-P., Li, S. X., Ma, R. C. W., Kong, A. P. S., Chan, M. H. M., Ho, C.-S., Li, A. M., & Wing, Y.-K. (2014). A community-based study on the association between insomnia and hypothalamic-pituitary-adrenal axis: Sex and pubertal influences. *The Journal of Clinical Endocrinology & Metabolism*, *99*(6), 2277–87. https://doi.org/10.1210/jc.2013-3728

Zhao, W., Van Someren, E. J. W., Li, C., Chen, X., Gui, W., Tian, Y., Liu, Y., & Lei, X. (2021). EEG spectral analysis in insomnia disorder: A systematic review and meta-analysis. *Sleep Medicine Reviews*, *59*, 101457. https://doi.org/10.1016/j. smrv.2021.101457

4

SLEEP DISRUPTION IN POSTTRAUMATIC STRESS DISORDER

Overview and relevant mechanisms

Laura D. Straus, Sara Rama, Kira Abirgas and Peter J. Colvonen

Acknowledgements and Disclosures

LDS is supported by the Department of Veterans Affairs Clinical Science Research and Development Award: IK2CX002032.

Posttraumatic stress disorder (PTSD) is a mental health condition present in approximately 7% of the general population (1), with particularly high prevalence in populations experiencing high rates of traumatic stress, including veterans (2), first responders (3) and victims of interpersonal violence (4). PTSD develops after an individual experiences a traumatic event, defined as an event during which a person has experienced, or been exposed to, actual or threatened death, serious injury, or sexual violence (5). While most individuals who experience trauma recover, some suffer consequences including reexperiencing symptoms (e.g., re-living trauma in the form of intrusive thoughts, becoming upset and physiologically aroused in response to trauma reminders), avoidance of trauma reminders and hyperarousal (e.g., constantly monitoring the environment for threats, showing an exaggerated startle response). PTSD is a chronic and disabling condition (6). Although evidence-based treatments have been developed for PTSD and are being implemented in various settings, including Veterans Affairs medical centres, not all patients respond to treatment (7). Understanding the factors underlying PTSD and/or interfering with treatment response is important to mitigate symptoms and ensure patients benefit from treatment.

Sleep disturbance is a near-universal symptom in PTSD. Sleep disruption in the form of insomnia and nightmares are included in the diagnostic criteria for

DOI: 10.4324/9781003296966-4

the disorder (5), and sleep difficulty is one of the most common self-reported complaints in those presenting with trauma-related concerns (8). Subjective and objective sleep problems often predate other trauma symptoms, are associated with worsening mental health concerns following a traumatic event, and often remain residually even after trauma-focused treatment (9–11). Thus, it has been suggested that sleep disturbance is a core feature in PTSD (12, 13). In addition, sleep disorders independently contribute to problematic substance use (14), suicidality (15) and poor quality of life (16) in individuals with trauma histories. Therefore, understanding relationships between sleep problems and PTSD will serve to inform interventions targeted toward improving sleep quality and reducing suffering in individuals with trauma histories. In this chapter, we will 1) summarise literature examining objective sleep disturbances in PTSD, 2) highlight variability of sleep as a key feature in this population, 3) review relevant literature suggesting Objective Sleep Apnoea (OSA) is a prevalent, challenging problem in this group and 4) discuss sleep disturbance and its relationship to fear and safety learning processes, which are mechanisms important in PTSD. We then discuss implications for assessment and treatment of sleep problems in PTSD and make suggestions for future research.

Objective sleep disruptions in PTSD

Given the prevalence of self-reported sleep problems in PTSD, a number of studies have investigated whether objective sleep stages are disrupted in this population. A 2007 meta-analysis of 20 polysomnography (PSG) studies comparing individuals with and without PTSD demonstrated increased stage 1 sleep and REM density in PTSD (17) and another meta-analysis showed greater objective wake after sleep onset in this patient group compared to controls (12). PTSD-positive participants have also been seen to experience decreased slow wave sleep (SWS) (17) as well as decreased sleep continuity (as evidenced by higher sleep latency, lower efficiency and more time awake after sleep onset) (18).

Despite these results, there may be demographic differences that moderate some of the objective sleep disruptions shown in PTSD compared to control participants. In the 2007 meta-analysis by Kobayashi and colleagues (17), age altered the size of the effect of PTSD on total sleep time (TST), SWS, REM sleep and REM latency. To complicate matters, differences seen in younger and older participants could be a reflection of alterations in PTSD based on age group or a reflection of the length of time elapsed since the index trauma. For example, research has suggested that nightmares may be more frequent in PTSD participants with more recent trauma, illustrating length of time since the index trauma has at least some effect on sleep symptoms (19). Sex may also moderate the relationship between sleep and PTSD. In studies with male participants, those with PTSD had shorter TST and stage 2 sleep, longer sleep onset latency and greater REM density. However, these differences were not observed in female participants. Additionally, participants in all-male studies are more likely to exclusively include veterans, while the participants in mixed gender studies were more likely to include civilians (19). Thus, the sex differences found in objective sleep in PTSD may be confounded by additional demographic variables.

Comorbid diagnoses may also moderate relationships between PTSD and objective sleep findings. PTSD diagnoses often co-occur with other psychiatric disorders, such as depression, anxiety and substance use disorders (SUD), all of which have been associated themselves with changes to sleep architecture as measured by PSG (20). Comorbid depression, which can be difficult to disentangle from PTSD (21), has been shown to be associated with a reduction in the sleep abnormalities seen in PTSD studies, moderating the effect of posttraumatic stress on TST, stage 1, SWS and length and density of REM sleep (17). By contrast, individuals with comorbid PTSD and SUD show more fragmentation in REM sleep compared to individuals with PTSD without SUD, suggesting higher arousal for these individuals (1). However, the direction of the effect is unclear – individuals with PTSD and SUD may have hyperarousal caused by use of substances, or the difference could be explained by pre-existing hyperarousal leading to use of substances to self-medicate.

Sleep Variability in PTSD

In addition to demographic factors and mental health comorbidities, another potential reason for discrepant findings regarding objective sleep disturbances in PTSD may be due to the *variability* of sleep in this population.

Sleep variability may present as within-group variability, when individuals in a group differ widely from each other as compared to other groups. Within-group variability can be statistically examined using bootstrapping and comparing standard deviations of variables between groups. In one study (22), a group of individuals with PTSD assessed their sleep for a week via sleep diaries and actigraphy, and results were compared with a separate group of individuals with non-comorbid insomnia and healthy controls. In terms of group means, only objective sleep efficiency was significantly worse with PTSD than with non-comorbid insomnia. However, the PTSD group displayed more interindividual variability on most measures, including subjective and objective TST, subjective and objective sleep efficiency, and subjective wake after sleep onset (see Figure 4.1). These results suggest when examining sleep, differences between individuals may be more prominent in PTSD than they are in other clinical groups – in this study, for example, some individuals showed sleep efficiencies comparable to healthy controls, while others showed sleep efficiencies that were much more severe than in the group with non-comorbid insomnia.

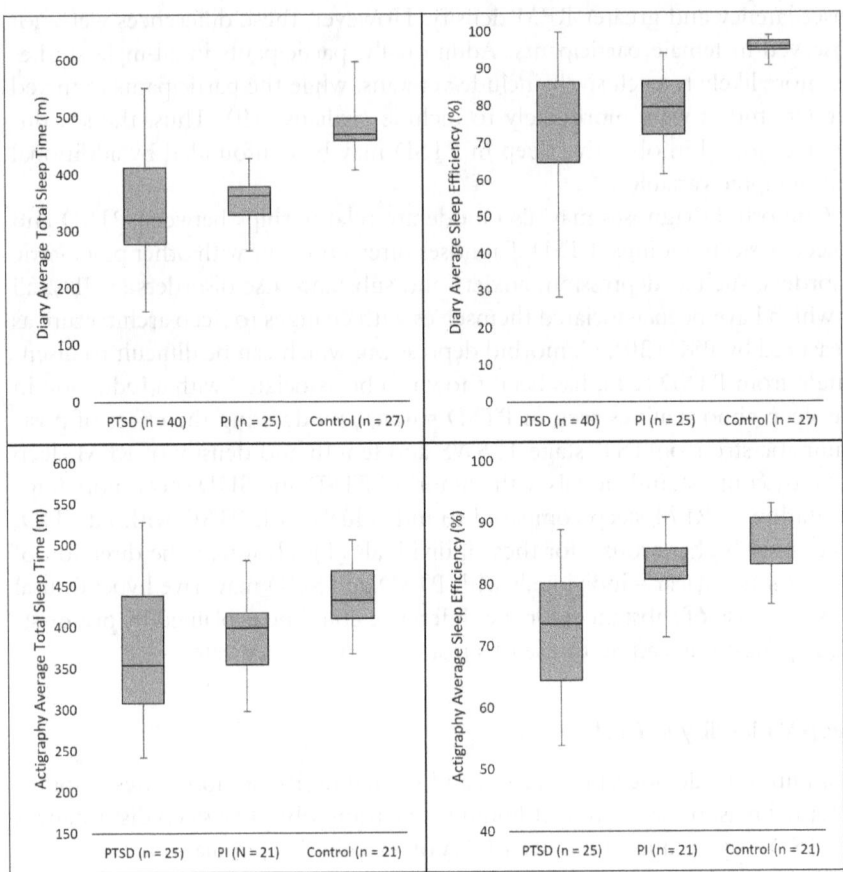

FIGURE 4.1 Display of within-group variability of several sleep variables in PTSD compared to primary insomnia and healthy controls.

Note: PTSD = posttraumatic stress disorder, PI = primary insomnia. Reproduced from Straus et al., *Journal of Traumatic Stress*, 2015

Another way in which sleep can be variable is via night-to-night variability within each individual. Night-to-night variability can be examined by using the mean squared of successive differences (MSSD), a measure similar to a standard deviation except that each score is compared to the score immediately preceding it rather than to the mean. In the study discussed above, sleep parameters from the PTSD group were compared to the group with non-comorbid insomnia and healthy controls in terms of night-to-night variability. Again, based on MSSD, the group with PTSD showed even more variability than the non-comorbid insomnia patients on subjective TST and objective sleep efficiency (Figure 4.2). These results suggest multi-timepoint measurements of sleep may be necessary for individuals with PTSD because a single

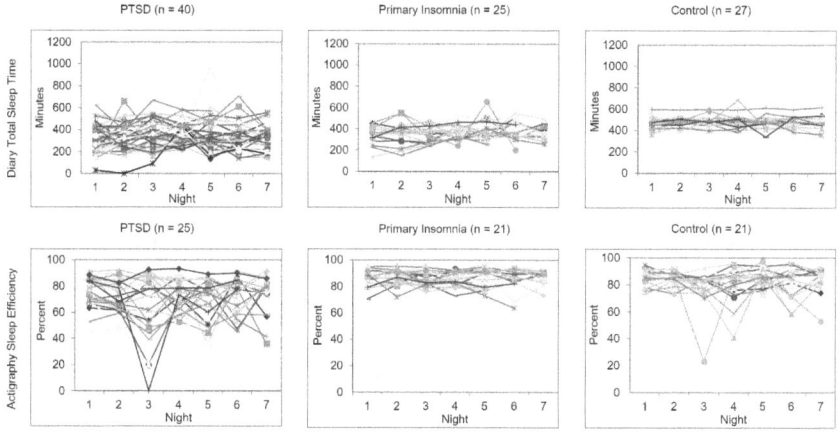

FIGURE 4.2   Night-to-night variability in PTSD. Display of night-to-night variability of diary total sleep time and actigraphy sleep efficiency in PTSD compared to primary insomnia and healthy controls, measured over one week. Each number on the X-axis represents one night of assessment, and each participant is represented by one line. Reproduced from Straus et al., *Journal of Traumatic Stress*, 2015.

measurement may not fully characterise sleep in a given individual. This also may explain discrepancies between studies measuring objective sleep in PTSD because assessment is highly variable depending on the particular night studied, and multi-night assessment can be difficult, especially when using objective measures such as PSG.

Comorbidity between PTSD and other sleep disorders

Another way in which sleep appears to be disrupted in PTSD is due to the considerable comorbidity between PTSD and Obstructive Sleep Apnoea (OSA), statistically the most common sleep disorder in the general population (23).

Sleep-disordered breathing is a spectrum that ranges from mild upper airway resistance (e.g., snoring) to severe OSA (24). OSA is defined by repeated episodes of hypopnoeas (ten seconds or more of shallow breathing) or apnoeas (stoppage in breathing). These disrupted breathing patterns lead to a decrease in oxygen during sleep that, in turn, leads to an brief awakening (cortical arousal), and thereby disrupting sleep continuity. The standard metric of OSA severity is the apnoea-hypopnoea index per hour (AHI) and ranges from mild (5–14 AHI per hour), moderate (15–29 AHI per hour), to severe (greater than 30 AHI per hour). OSA is diagnosed using objective overnight testing. Additionally, self-report screeners are almost universally used before overnight testing is ordered. Measures commonly used to screen for OSA are the Berlin

Questionnaire (Berlin), the STOP-BANG, and the Epworth Sleepiness Scale (ESS). The Berlin and STOP-BANG measure risk factors for OSA (e.g., high blood pressure, old age, male gender and high body mass index, BMI) (25) and the ESS measures daytime sleepiness symptoms (e.g., likelihood of falling asleep while driving).

Rates of OSA in PTSD specialty clinics are suggested to be high (26) but the exact prevalence is unknown. There are three reasons for this. First, screening for OSA is not a part of standard clinical care in PTSD specialty clinics, and as such, has low objective screening referrals to pulmonary sleep clinics (27). Second, the daytime symptoms of OSA, such as fatigue, memory problems, trouble concentrating, and difficulty with emotional coping, have significant overlap with PTSD symptoms and are often associated with the 'primary disorder'. Third, there is increasing evidence that the classic predictors of OSA, such as BMI above 30, high blood pressure, and older age, may not apply to younger veterans with PTSD. As such, the most accessible self-report screening tools (e.g., Berlin) that rely on classic predictors have false negatives in approximately 25% of individuals with PTSD and OSA due to an atypical OSA presentation (28).

Two recent studies found 67.3–69.2% of younger veterans (mean age = 33.4–35.1 years) with PTSD and lower BMI (BMI = 19.1–28.9) than typically seen in 'classic' OSA were at high risk of OSA (26, 29). Similarly, in a recent PSG study comparing Iranian veterans with and without PTSD, Rezaeitalab et al. (30) found AHI was higher and BMI lower in the PTSD group compared to the non-PTSD group, and that AHI was unrelated to BMI in the group with PTSD. OSA is increasingly recognised as a multifactorial disorder; i.e., different people have OSA for different reasons (31–34). Although an underlying anatomical predisposition (collapsibility of the upper airway) is required (35, 36), other factors can be important as well: a low respiratory arousal threshold (waking up too easily from sleep in response to small respiratory cues) is one hypothesised mechanism for atypical OSA. It has been suggested that the chronic daytime hyperarousal of PTSD may manifest as a low arousal threshold leading to a lower threshold for waking from sleep, causing fragmentation and increased AHI (37). This hypothesis has been supported by findings from El-Sohl and colleagues, who estimated participants' arousal thresholds (38) and found that 55% of veterans with PTSD and OSA had a low arousal threshold (39). These 'atypical' or alternative endotypes of OSA may also help explain why approximately 80–90% of individuals with OSA and PTSD remain undiagnosed (27, 40), and why OSA screeners that rely on age, blood pressure, and BMI may miss many individuals with comorbid PTSD (28).

The rates of undiagnosed and untreated OSA are concerning because PTSD treatment efficacy is hindered by untreated OSA (41, 42). Prolonged exposure treatment is considered the gold-standard, evidence-based treatment for PTSD. Reist and colleagues found participants undergoing prolonged exposure without sleep-disordered breathing had a large drop in the PTSD checklist (PCL) scores, with a 28.25 point reduction that is expected following treatment (41). However, among individuals with untreated sleep-disordered breathing, there was only a small PCL score reduction of 7.17 points following prolonged exposure therapy. Similarly, a retrospective study of individuals who had completed cognitive processing therapy at a VA centre found that those with OSA (n = 69) showed less symptom improvement than those without OSA (n = 276) (42). Finally, untreated OSA was shown to negatively impact ketamine treatment for Veterans who presented with unresponsive pharmacological depression and PTSD (43). Taken together, this suggests that treating OSA may be a necessary first step in treating PTSD.

The gold-standard treatment for OSA is positive airway pressure (PAP), with the most common prescription being to use the machine whenever sleeping or napping. PAP use has been shown to improve symptoms of daytime sleepiness and health-related quality of life (27). Furthermore, increased PAP use is associated with decreased nightmare severity and less daytime sleepiness (44). Likewise, frequent PAP use has been found to decrease the frequency of nightmares from 10.23 to 5.26 nightmares per week (43). PAP use has also been found to be associated with decreases in PTSD severity at 12 weeks (45) and 6 months (46, 47). Colvonen and colleagues (47) found that, among veterans with PTSD and OSA, the high PAP adherent group showed a 14.36-point decrease on the PCL-S, while the low adherent group only averaged a 3.66-point decrease.

Unfortunately, PAP adherence rates among individuals with PTSD are low, which decreases the efficacy of PAP therapy. (48). Rates of PAP adherence in individuals with PTSD range from 39% to 81% depending upon the definition of adherence (46, 49, 50). Among veterans with PTSD, higher levels of reexperiencing and hyperarousal, but not avoidance, predicted lower PAP use (47). This suggests that individuals with PTSD, especially those displaying significant reexperiencing or hyperarousal symptoms, require extra clinical attention at the time of PAP initiation. Using PAP desensitisation protocols (e.g., wearing PAP while watching TV, practising mindfulness while wearing PAP) and instruction on how to successfully use the RAMP function (decreasing initial PAP pressure for 5–45 minutes while Veteran falls asleep or recovers from a nightmare) may have significant effects on adherence and treatment outcomes. However, no studies to date have systematically examined desensitisation protocols among individuals with PTSD receiving PAP treatment.

The high prevalence rates of comorbid OSA and PTSD are concerning, especially in the light of low PAP adherence rates and the threat unmanaged OSA poses on the success of a variety of PTSD treatments. As OSA requires direct intervention separate from PTSD treatment, assessing and addressing OSA before or shortly after PTSD treatment is initiated may prove vital to improving treatment efficacy for both disorders. However, clinical guidelines are needed for screening for sleep disorders comorbid with PTSD given that currently

OSA is rarely identified due to the overlap with PTSD and insomnia symptoms (4). The sleep and PTSD treatment decision tree proposed by Colvonen and colleagues suggests a comprehensive sleep assessment for both OSA and insomnia in conjunction with coordinated treatment planning (51). Creating effective screening tools for OSA in the context of PTSD, and increasing screening for and treatment of OSA, is a critical healthcare issue for improving functioning and quality of life among veterans as research suggests that sleep improvements would have a positive effect on fear learning (discussed below), an important mechanism of PTSD treatment. In general, more literature examining change in AHI over the course of PTSD treatment and efficacy of coordinating OSA + PTSD treatments for patients with both disorders is needed.

Sleep and memory processes in PTSD

Sleep and fear extinction learning and memory

A major reason for studying sleep disruptions and sleep-related comorbidities in PTSD is that sleep disturbances influence other mechanisms relevant to cardinal symptoms of the disorder. Clinically, patients with PTSD experience a heightened fear response to cues associated with a traumatic event even though these cues in themselves may not be inherently dangerous. For example, a combat veteran may experience fear when seeing debris on the side of the road while driving, because this cue signalled danger (the presence of an explosive device) when on deployment. These individuals also often show this fear response long after the trauma, indicating difficulty with extinction learning, i.e., learning that these neutral cues no longer indicate threat. Research examining these mechanisms often makes use of Pavlovian fear conditioning and extinction learning paradigms (52, 53) (see Figure 4.3 for a schematic illustrating a fear conditioning experiment). In these studies, participants undergo a fear conditioning session, during which they are presented with neutral cues (e.g., coloured circles), some of which are repeatedly paired with a noxious stimulus, such as a shock to the wrist or puff of air to the throat. After the session, participants then undergo an extinction learning session, during which the threat cues are presented repeatedly without the noxious stimulus. This provides an opportunity for extinction learning, i.e., learning that the neutral stimulus no longer predicts objective threat. At subsequent extinction recall sessions, participants are again presented with the threat cue from the fear conditioning session, providing an opportunity for researchers to learn whether extinction learning was retained or if the fear response returns.

Various studies have manipulated sleep at various points in fear conditioning experiments and have consistently linked disrupted sleep with impaired extinction processes. While most of these studies have been conducted in healthy control participants rather than in individuals with PTSD, several have shown that sleep disruption impairs extinction learning (54, 55), retention (56, 57),

Fear/safety learning
(Session 1)

Extinction learning
(Session 2)

Extinction recall
(Session 3)

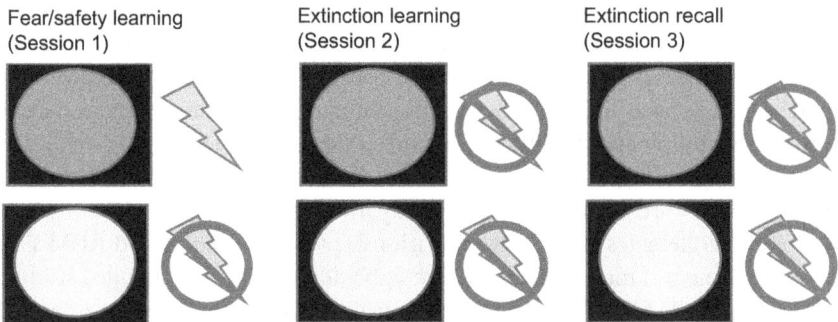

FIGURE 4.3   Schematic of fear conditioning experiments. Sequence depicting lab-based experiments assessing fear conditioning, safety learning and extinction processes. In Session 1, participants are presented with two neutral cues (blue and yellow circles), one of which is paired repeatedly with a noxious stimulus (e.g., shock to the wrist, represented by lightning bolt). Thus, one stimulus (blue circle) becomes a threat cue. The other stimulus (yellow circle) is never presented with the noxious cue and becomes a safety signal. At subsequent sessions, both cues are again presented without the noxious stimulus, offering the opportunity for participants to experience extinction learning at Session 2 and recall at Session 3.

and/or generalisation (58). Additionally, sleep deprivation has been shown to interfere with the neural correlates of extinction processes (59). REM consolidation seems especially critical given that individuals who had REM sleep during a nap showed better extinction recall than those who did not (60), and another study showed that selectively depriving individuals of REM sleep interfered with extinction recall (61). Taken together, these findings demonstrate how disrupted sleep, especially REM sleep, impairs these processes and thus may influence clinical symptoms, especially fear-related symptoms.

Sleep and safety signal learning and memory

In addition to extinction processes, safety learning is another critical process involving the ability to differentiate cues that predict threat versus safety. This is a critical feature in daytime PTSD symptoms and has been proposed as PTSD-specific (62). For example, a combat veteran with PTSD may experience a heightened fear response to a cue that had previously signalled threat (e.g., helicopter sound), even in the presence of many safety cues (e.g., surrounded by family and friends, far away from the combat theatre). In laboratory studies, safety signal learning is operationalised in fear conditioning paradigms by examining response to the neutral cues (e.g., yellow circles) that are never presented along with the threat stimulus (e.g., shock; see Figure 4.3).

Relatively fewer studies have examined the role of sleep in safety signal learning. However, one study in healthy control participants (63) showed safety signal learning was associated with more consolidated subsequent REM sleep, which, in turn, was linked to better discrimination between threat and safety signals the following day. Richards and colleagues (64) recently replicated these findings using a nap paradigm in individuals with trauma histories. To date, we are aware of one additional study involving participants with PTSD – Straus and colleagues (65) showed relationships between increased REM sleep consolidation and more efficient safety signal learning in a small pilot study of veterans with PTSD. Though to our knowledge these are the only three studies to date that have examined relationships between sleep architecture and safety learning, they provide strong evidence of the importance of REM sleep for safety learning in addition to fear processes.

Summary and implications

Based on the prevailing research, sleep is disrupted and variable in PTSD. Furthermore, PTSD is frequently comorbid with other disorders such as OSA, which may result in additional sleep disruptions in this population, especially due to specific challenges to diagnosing and/or treating OSA in PTSD. Additionally, poor sleep is associated with impaired extinction and safety signal learning processes, suggesting poor sleep may maintain or worsen other features of PTSD.

These findings have considerable implications when considering clinical assessment of sleep disorders in PTSD. The large within-group and night-to-night variability of sleep in PTSD suggests an individualised approach and multi-night assessment are critical when assessing sleep problems (22). Additionally, the high rates of OSA in PTSD, particularly in patients without other 'classic' OSA predictors (e.g., BMI > 30, high blood pressure, older age) or who score negative on common OSA screeners, suggest many cases of OSA may be overlooked and thus broad-based OSA screening is indicated, especially in specialty PTSD clinics.

This body of research also has implications for treating sleep disorders in PTSD. REM sleep in particular appears to impair extinction and safety signal processes, suggesting poor sleep likely enhances daytime PTSD symptoms. Given the high rates of OSA in PTSD, OSA may compound the contribution of poor sleep to impairments in these mechanisms – REM sleep appears to be especially critical for fear extinction and safety learning and REM is the sleep stage during which breathing events are most likely to occur (66). Notably, gold-standard psychotherapeutic interventions for PTSD require successful extinction and safety signal learning. For example, in prolonged exposure for PTSD, patients repeatedly expose themselves to feared cues in a safe environment, allowing safety and extinction learning to occur. Poor sleep appears to interfere with these processes, suggesting sequencing sleep treatment first for those who need it (especially OSA treatment) may be critical for ensuring patients have adequate response to treatments like prolonged exposure.

Based on the findings we discuss above, we have several suggestions for future research. As noted above, research should consider demographic factors and comorbidities when searching for objective sleep disruptions in PTSD. Additionally, given the variability of sleep between individuals and from night to night in PTSD, large multi-night studies may be needed to tease out sleep features that may have particular importance in this population. Although large studies of sleep involving multiple timepoints and objective sleep measures have historically been difficult and expensive to implement, emergence of new wearable technologies may facilitate this research, especially with devices that are easy to use remotely. Second, we have summarised recent studies suggesting OSA is especially prevalent in PTSD, and we have highlighted challenges to adherence to PAP treatment for this patient group. Future studies of PTSD should trial OSA-specific treatments, such as PAP desensitisation. Third, we have summarised recent neuroscience research, conducted mostly in healthy control participants, suggesting impaired sleep interferes with fear extinction and safety learning processes. The implication is that sleep treatment may be especially important to reduce other daytime symptoms of PTSD and/or facilitate treatment response. Future studies should focus on testing these hypotheses directly, especially in clinical populations.

References

1 Kessler, R. C., Berglund, P., Demler, O., Jin, R., Merikangas, K. R., & Walters, E. E. (2005). Lifetime prevalence and age-of-onset distributions of DSM-IV disorders in the National Comorbidity Survey Replication. *Archives of General Psychiatry, 62*(6), 593–602.

2 Blake, D. D., Keane, T. M., Wine, P. R., Mora, C., Taylor, K. L., & Lyons, J. A. (1990). Prevalence of PTSD symptoms in combat veterans seeking medical treatment. *Journal of Traumatic Stress, 3*, 15–27.

3 Berger, W., Coutinho, E. S. F., Figueira, I., Marques-Portella, C., Luz, M. P., Neylan, T. C., . . . & Mendlowicz, M. V. (2012). Rescuers at risk: a systematic review and meta-regression analysis of the worldwide current prevalence and correlates of PTSD in rescue workers. *Social Psychiatry and Psychiatric Epidemiology, 47*, 1001–11.

4 Nathanson, A. M., Shorey, R. C., Tirone, V., & Rhatigan, D. L. (2012). The prevalence of mental health disorders in a community sample of female victims of intimate partner violence. *Partner Abuse, 3*(1), 59–75.

5 American Psychiatric Association. (2022). *Diagnostic and statistical manual of mental disorders* (5th ed., text rev.). https://doi.org/10.1176/appi. books.9780890425787

6 Davidson, J. R., Stein, D. J., Shalev, A. Y., & Yehuda, R. (2004). Posttraumatic stress disorder: acquisition, recognition, course, and treatment. *The Journal of Neuropsychiatry and Clinical Neurosciences, 16*(2), 135–47.

7 Maguen, S., Li, Y., Madden, E., Seal, K. H., Neylan, T. C., Patterson, O. V., . . . & Shiner, B. (2019). Factors associated with completing evidence-based psychotherapy

for PTSD among veterans in a national healthcare system. *Psychiatry Research, 274,* 112–28.

8 McLay, R. N., Klam, W. P., & Volkert, S. L. (2010). Insomnia is the most commonly reported symptom and predicts other symptoms of post-traumatic stress disorder in US service members returning from military deployments. *Military Medicine, 175*(10), 759–62.

9 Neylan, T. C., Kessler, R. C., Ressler, K. J., Clifford, G., Beaudoin, F. L., An, X., . . . & McLean, S. A. (2021). Prior sleep problems and adverse post-traumatic neuropsychiatric sequelae of motor vehicle collision in the AURORA study. *Sleep, 44*(3), zsaa200.

10 Straus, L. D., An, X., Ji, Y., McLean, S. A., Neylan, T. C., Cakmak, A. S., . . . & AURORA Study Group. (2023). Utility of wrist-wearable data for assessing pain, sleep, and anxiety outcomes after traumatic stress exposure. *JAMA Psychiatry, 80*(3), 220–9.

11 Gutner, C. A., Casement, M. D., Gilbert, K. S., & Resick, P. A. (2013). Change in sleep symptoms across cognitive processing therapy and prolonged exposure: a longitudinal perspective. *Behaviour Research and Therapy, 51*(12), 817–22.

12 Germain, A. (2013). Sleep disturbances as the hallmark of PTSD: where are we now? *American Journal of Psychiatry, 170*(4), 372–82.

13 Miller, K. E., Brownlow, J. A., Woodward, S., & Gehrman, P. R. (2017). Sleep and dreaming in posttraumatic stress disorder. *Current Psychiatry Reports, 19,* 1–10.

14 Nishith, P., Resick, P. A., & Mueser, K. T. (2001). Sleep difficulties and alcohol use motives in female rape victims with posttraumatic stress disorder. *Journal of Traumatic Stress: Official Publication of the International Society for Traumatic Stress Studies, 14*(3), 469–79.

15 Porras-Segovia, A., Perez-Rodriguez, M. M., López-Esteban, P., Courtet, P., López-Castromán, J., Cervilla, J. A., & Baca-García, E. (2019). Contribution of sleep deprivation to suicidal behaviour: a systematic review. *Sleep Medicine Reviews, 44,* 37–47.

16 Krakow, B., Melendrez, D., Johnston, L., Warner, T. D., Clark, J. O., Pacheco, M., . . . & Schrader, R. (2002). Sleep-disordered breathing, psychiatric distress, and quality of life impairment in sexual assault survivors. *The Journal of Nervous and Mental Disease, 190*(7), 442–52.

17 Kobayashi, I., Boarts, J. M., & Delahanty, D. L. (2007). Polysomnographically measured sleep abnormalities in PTSD: a meta-analytic review. *Psychophysiology, 44*(4), 660–9.

18 Baglioni, C., Nanovska, S., Regen, W., Spiegelhalder, K., Feige, B., Nissen, C., . . . & Riemann, D. (2016). Sleep and mental disorders: a meta-analysis of polysomnographic research. *Psychological Bulletin, 142*(9), 969.

19 Richards, A., Kanady, J. C., & Neylan, T. C. (2020). Sleep disturbance in PTSD and other anxiety-related disorders: an updated review of clinical features, physiological characteristics, and psychological and neurobiological mechanisms. *Neuropsychopharmacology, 45*(1), 55–73.

20 Zhang, Y., Ren, R., Sanford, L. D., Yang, L., Zhou, J., Zhang, J., . . . & Tang, X. (2019). Sleep in posttraumatic stress disorder: a systematic review and meta-analysis of polysomnographic findings. *Sleep Medicine Reviews, 48,* 101210.

21 O'Donnell, M. L., Creamer, M., & Pattison, P. (2004). Posttraumatic stress disorder and depression following trauma: understanding comorbidity. *American Journal of Psychiatry, 161*(8), 1390–6.

22 Straus, L. D., Drummond, S. P., Nappi, C. M., Jenkins, M. M., & Norman, S. B. (2015). Sleep variability in military-related PTSD: a comparison to primary insomnia and healthy controls. *Journal of Traumatic Stress, 28*(1), 8–16.

23 Franklin, K. A., & Lindberg, E. (2015). Obstructive sleep apnea is a common disorder in the population – a review on the epidemiology of sleep apnea. *Journal of Thoracic Disease, 7*(8), 1311.

24 Schwab, R. J., Goldberg, A. N., & Pack, A. L. (1998). Sleep apnea syndromes. *Fishman's Pulmonary Diseases and Disorders. New York: McGraw-Hill Book Company, 1617*, 37.
25 Peppard, P. E., Young, T., Palta, M., Dempsey, J., Skatrud, J. (2000). Longitudinal study of moderate weight change and sleep-disordered breathing. *Jama, 284*(23), 3015–21.
26 Colvonen, P. J., Masino, T., Drummond, S. P., Myers, U. S., Angkaw, A. C., & Norman, S. B. (2015). Obstructive sleep apnea and posttraumatic stress disorder among OEF/OIF/OND veterans. *Journal of Clinical Sleep Medicine, 11*(5), 513–18.
27 Colvonen, P. J., Straus, L. D., Stepnowsky, C., McCarthy, M. J., Goldstein, L. A., Norman, S. B. (2018). Recent advancements in treating sleep disorders in co-occurring PTSD. *Current Psychiatry Reports, 20*(7), 48.
28 Lyons, R., Barbir, L., Norman, S. B., Owens, R., & Colvonen, P. J. (2020). Examining the association between subjective and objective measures of obstructive sleep apnea risk in veterans with posttraumatic stress disorder and insomnia. *Journal of Clinical Sleep Medicine, 18*(1), 67–73.
29 Williams, S. G., Collen, J., Orr, N., Holley, A. B., Lettieri, C. J. (2015). Sleep disorders in combat-related PTSD. *Sleep and Breathing, 19*(1), 175–82.
30 Rezaeitalab, F., Mokhber, N., Ravanshad, Y., Saberi, S., & Rezaeetalab, F. (2018). Different polysomnographic patterns in military veterans with obstructive sleep apnea in those with and without post-traumatic stress disorder. *Sleep and Breathing, 22*, 17–22.
31 Martinez-Garcia, M. A., Campos-Rodriguez, F., Barbé, F., Gozal, D., & Agustí, A. (2019). Precision medicine in obstructive sleep apnoea. *The Lancet Respiratory Medicine, 7*(5), 456–64.
32 Mazzotti, D. R., Lim, D. C., Sutherland, K., Bittencourt, L., Mindel, J. W., Magalang, U., . . . & Penzel, T. (2018). Opportunities for utilizing polysomnography signals to characterize obstructive sleep apnea subtypes and severity. *Physiological Measurement, 39*(9), 09TR01.
33 Pack, A. I. (2019). Further development of P4 approach to obstructive sleep apnea. *Sleep Medicine Clinics, 14*(3), 379–89.
34 Zinchuk, A., & Yaggi, H. K. (2020). Phenotypic subtypes of OSA: a challenge and opportunity for precision medicine. *Chest, 157*(2), 403–20.
35 Schwab, R. J., Leinwand, S. E., Bearn, C. B., Maislin, G., Rao, R. B., Nagaraja, A., . . . & Keenan, B. T. (2017). Digital morphometrics: a new upper airway phenotyping paradigm in OSA. *Chest, 152*(2), 330–42.
36 Schwartz, A. R., Rowley, J. A., Thut, D. C., Permutt, S., & Smith, P. L. (1996). Structural basis for alterations in upper airway collapsibility. *Sleep, 19*(suppl_10), 184–8.
37 Lettieri, C. J., Collen, J. F., & Williams, S. G. (2017). Challenges in the management of sleep apnea and PTSD: is the low arousal threshold an unrealized target? *Journal of Clinical Sleep Medicine, 13*(6), 845–6.
38 Edwards, B. A., Eckert, D. J., McSharry, D. G., Sands, S. A., Desai, A., Kehlmann, G., . . . & Malhotra, A. (2014). Clinical predictors of the respiratory arousal threshold in patients with obstructive sleep apnea. *American Journal of Respiratory and Critical Care Medicine, 190*(11), 1293–300.
39 El-Solh, A. A., Lawson, Y., & Wilding, G. E. (2021). Impact of low arousal threshold on treatment of obstructive sleep apnea in patients with post-traumatic stress disorder. *Sleep and Breathing, 25*, 597–604.
40 Alexander, M., Ray, M. A., Hébert, J. R., Youngstedt, S. D., Zhang, H., Steck, S. E., . . . & Burch, J. B. (2016). The national veteran sleep disorder study: descriptive epidemiology and secular trends, 2000–2010. *Sleep, 39*(7), 1399–410.

41 Reist, C., Gory, A., & Hollifield, M. (2017). Sleep-disordered breathing impact on efficacy of prolonged exposure therapy for posttraumatic stress disorder. *Journal of Traumatic Stress, 30*(2), 186–9.

42 Mesa, F., Dickstein, B. D., Wooten, V. D., & Chard, K. M. (2017). Response to cognitive processing therapy in veterans with and without obstructive sleep apnea. *Journal of Traumatic Stress, 30*(6), 646–55.

43 Tamanna, S., Parker, J. D., Lyons, J., & Ullah, M. I. (2014). The effect of continuous positive air pressure (CPAP) on nightmares in patients with posttraumatic stress disorder (PTSD) and obstructive sleep apnea (OSA). *Journal of Clinical Sleep Medicine, 10*(6), 631–6.

44 El-Solh, A. A., Ayyar, L., Akinnusi, M., Relia, S., & Akinnusi, O. (2010). Positive airway pressure adherence in veterans with posttraumatic stress disorder. *Sleep, 33*(11), 1495–500.

45 El-Solh, A. A., Vermont, L., Homish, G. G., & Kufel, T. (2017). The effect of continuous positive airway pressure on post-traumatic stress disorder symptoms in veterans with post-traumatic stress disorder and obstructive sleep apnea: a prospective study. *Sleep Medicine, 33*, 145–50.

46 Orr, J. E., Smales, C., Alexander, T. H., Stepnowsky, C., Pillar, G., Malhotra, A., & Sarmiento, K. F. (2017). Treatment of OSA with CPAP is associated with improvement in PTSD symptoms among veterans. *Journal of Clinical Sleep Medicine, 13*(1), 57–63.

47 Colvonen, P. J., Goldstein, L. A., & Sarmiento, K. F. (2023). Examining the bidirectional relationship between PTSD symptom clusters and PAP adherence. *Journal of Clinical Sleep Medicine*, jcsm-10430.

48 Zhang, Y., Weed, J. G., Ren, R., Tang, X., & Zhang, W. (2017). Prevalence of obstructive sleep apnea in patients with posttraumatic stress disorder and its impact on adherence to continuous positive airway pressure therapy: a meta-analysis. *Sleep Medicine, 36*, 125–32.

49 Collen, J. F., Lettieri, C. J., & Hoffman, M. (2012). The impact of posttraumatic stress disorder on CPAP adherence in patients with obstructive sleep apnea. *Journal of Clinical Sleep Medicine, 8*(6), 667–72.

50 Ullah, M. I., Campbell, D. G., Bhagat, R., Lyons, J. A., & Tamanna, S. (2017). Improving PTSD symptoms and preventing progression of subclinical PTSD to an overt disorder by treating comorbid OSA with CPAP. *Journal of Clinical Sleep Medicine, 13*(10), 1191–8.

51 Flemons, W. W., Douglas, N. J., Kuna, S. T., Rodenstein, D. O., & Wheatley, J. (2004). Access to diagnosis and treatment of patients with suspected sleep apnea. *American Journal of Respiratory and Critical Care Medicine, 169*(6), 668–72.

52 Mahan, A. L., & Ressler, K. J. (2012). Fear conditioning, synaptic plasticity and the amygdala: implications for posttraumatic stress disorder. *Trends in Neurosciences, 35*(1), 24–35.

53 Careaga, M. B. L., Girardi, C. E. N., & Suchecki, D. (2016). Understanding posttraumatic stress disorder through fear conditioning, extinction and reconsolidation. *Neuroscience & Biobehavioral Reviews, 71*, 48–57.

54 Sturm, A., Czisch, M., & Spoormaker, V. I. (2013). Effects of unconditioned stimulus intensity and fear extinction on subsequent sleep architecture in an afternoon nap. *Journal of Sleep Research, 22*(6), 648–55.

55 Lerner, I., Lupkin, S. M., Sinha, N., Tsai, A., & Gluck, M. A. (2017). Baseline levels of rapid eye movement sleep may protect against excessive activity in fear-related neural circuitry. *Journal of Neuroscience, 37*(46), 11233–44.

56 Straus, L. D., Acheson, D. T., Risbrough, V. B., & Drummond, S. P. (2017). Sleep deprivation disrupts recall of conditioned fear extinction. *Biological Psychiatry: Cognitive Neuroscience and Neuroimaging, 2*(2), 123–9.

57 Bottary, R., Seo, J., Daffre, C., Gazecki, S., Moore, K. N., Kopotiyenko, K., . . . & Pace-Schott, E. F. (2020). Fear extinction memory is negatively associated with REM sleep in insomnia disorder. *Sleep*, *43*(7), zsaa007.

58 Pace-Schott, E. F., Milad, M. R., Orr, S. P., Rauch, S. L., Stickgold, R., & Pitman, R. K. (2009). Sleep promotes generalization of extinction of conditioned fear. *Sleep*, *32*(1), 19–26.

59 Seo, J., Pace-Schott, E. F., Milad, M. R., Song, H., & Germain, A. (2021). Partial and total sleep deprivation interferes with neural correlates of consolidation of fear extinction memory. *Biological Psychiatry: Cognitive Neuroscience and Neuroimaging*, *6*(3), 299–309.

60 Spoormaker, V. I., Sturm, A., Andrade, K. C., Schröter, M. S., Goya-Maldonado, R., Holsboer, F., . . . & Czisch, M. (2010). The neural correlates and temporal sequence of the relationship between shock exposure, disturbed sleep and impaired consolidation of fear extinction. *Journal of Psychiatric Research*, *44*(16), 1121–8.

61 Spoormaker, V. I., Gvozdanovic, G. A., Sämann, P. G., & Czisch, M. (2014). Ventromedial prefrontal cortex activity and rapid eye movement sleep are associated with subsequent fear expression in human subjects. *Experimental Brain Research*, *232*, 1547–54.

62 Jovanovic, T., Kazama, A., Bachevalier, J., & Davis, M. (2012). Impaired safety signal learning may be a biomarker of PTSD. *Neuropharmacology*, *62*(2), 695–704.

63 Marshall, A. J., Acheson, D. T., Risbrough, V. B., Straus, L. D., & Drummond, S. P. (2014). Fear conditioning, safety learning, and sleep in humans. *Journal of Neuroscience*, *34*(35), 11754–60.

64 Richards, A., Inslicht, S. S., Yack, L. M., Metzler, T. J., Russell Huie, J., Straus, L. D., . . . & Neylan, T. C. (2022). The relationship of fear-potentiated startle and polysomnography-measured sleep in trauma-exposed men and women with and without PTSD: testing REM sleep effects and exploring the roles of an integrative measure of sleep, PTSD symptoms, and biological sex. *Sleep*, *45*(1), zsab271.

65 Straus, L. D., Norman, S. B., Risbrough, V. B., Acheson, D. T., & Drummond, S. P. (2018). REM sleep and safety signal learning in posttraumatic stress disorder: a preliminary study in military veterans. *Neurobiology of Stress*, *9*, 22–8.

66 Alzoubaidi, M., & Mokhlesi, B. (2016). Obstructive sleep apnea during REM sleep: clinical relevance and therapeutic implications. *Current Opinion in Pulmonary Medicine*, *22*(6), 545.

5

SLEEP DISTURBANCES AND DISORDERS – A RISK FACTOR FOR COGNITIVE DECLINE AND DEMENTIA

Aaron Lam, Camilla Hoyos, Craig Phillips and Sharon L. Naismith

Overview of dementia and mild cognitive impairment (MCI)

Dementia is an umbrella term referring to a group of neurodegenerative conditions that cause cognitive impairment and impaired daily functioning. Worldwide, approximately 55 million people have dementia and it is a leading cause of mortality (1). The most common forms of dementia are Alzheimer's Disease (AD) and vascular dementia (VaD), which account for up to 70% (2) of cases and 14.5% (3) of diagnosed cases, respectively. Other major forms of dementia include Dementia with Lewy Bodies and frontotemporal dementia, which account for up to 5.4% (3) and 4% (4) of cases, respectively. In a minority of cases, other conditions, such as Parkinson's disease, multiple sclerosis and Huntington's disease may cause dementia. While there is some overlap in clinical presentations, each form of dementia is pathologically distinct.

This chapter will focus on AD and VaD, which often co-occur, collectively accounting for up to 80% of cases (5). A pathological hallmark of AD is beta-amyloid (Aβ) plaques, misfolded proteins in the brain that are believed to lead to synaptic dysfunction, neural dysconnectivity and neuronal death (5). Another critical protein in AD pathology is tau neurofibrillary tangles. Tau proteins are found in nerve cells and help stabilise nerve cell structures, however in AD, hyper-phosphorylation of tau leads to aggregation of neurofibrillary tangles. These neurofibrillary tangles are thought to contribute to neuronal death and cognitive decline associated with AD (6). In contrast, VaD is characterised by arteriosclerosis, lacunar infarcts and diffuse white matter changes (5). As implied, VaD is linked to vascular risk factors such as hypertension, hypercholesterolaemia, heart disease, smoking and diabetes.

DOI: 10.4324/9781003296966-5

There are several groups of individuals that are considered to be 'at-risk' of dementia, including those with:

1) *Subjective Cognitive Complaints (SCC)* – cognitive concerns but no detectable objective cognitive impairment on neuropsychological testing. A person with SCC has almost four times the likelihood of conversion to either MCI or dementia than a person without (7).
2) *Preclinical AD* – biomarker evidence of Aβ on positron emission tomography (PET) or cerebrospinal fluid (CSF) (8). Up to 18% of individuals without cognitive impairment but are Aβ-positive convert to MCI or dementia over a 15-month period (9).
3) *Mild Cognitive Impairment (MCI)* – objective cognitive impairment (typically 1.5 standard deviation decline) on formal neuropsychological testing, but with generally intact daily functioning (thus not meet criteria for dementia). Around 9.4% (10) will convert to dementia each year, with approximately 45% progressing over a five-year period.

While there have been some promising recent advances in disease-modifying therapies (e.g., Lecanemab which reduces amyloid burden (11)), there is currently no known cure for dementia. Importantly, the pathological changes leading to dementia accumulate over one to two decades (12), prompting efforts to consider risk reduction from midlife. Indeed, a recent Lancet Commission report calculated that approximately 40% of the risk for dementia (notably AD and VaD) can be attributed to modifiable risk factors including less education, hearing loss, traumatic brain injury, hypertension, excessive alcohol consumption, obesity, smoking, depression, social isolation, physical inactivity, diabetes and air pollution (13). Hence, there is considerable interest in understanding the mechanisms by which these modifiable risk factors might contribute to dementia pathology. Such insights could inform the timing and delivery of interventions that could optimise cognition in critical at-risk periods.

Notably, the Lancet Commission report did not include sleep disturbance as a risk factor. Instead, it was outlined as a potential modifiable factor that required more research into mechanisms and large-scale clinical trials (13). The subsequent sections of the chapter will summarise the critical role of sleep for brain health and cognition, and the status of epidemiological, mechanistic and clinical trial knowledge in the field.

Table 4.1 outlines the sleep disturbances and formal sleep disorders that have been studied in the context of ageing and dementia.

Importance of sleep for brain health and cognition

Sleep is critical for brain health, mood and memory. For instance, slow-wave activity (0.5–4.5 Hz) during slow wave sleep appears to promote the

TABLE 4.1 Sleep disorders that are commonly studied in ageing and dementia.

Sleep disturbance	Characterisation
Poor sleep quality and general sleep disturbance	Self-reported decrease in sleep duration and quality, or unrefreshing sleep. Using actigraphy, this can be quantified by reduced sleep efficiency, increased wake after sleep onset or sleep fragmentation. Using polysomnography, this can be associated with reduced total sleep time, wake after sleep onset or reduced rapid eye movement (REM) sleep or slow-wave sleep (SWS).
Circadian alterations/ misalignment	May present as circadian advance or delay, or inappropriately timed sleep/wake, appetite or other circadian behaviours relative to sleep. Ideally measured via gold-standard measures of melatonin, but actigraphy is often used to measure sleep-wake rhythms.
Sleep disorder	Characterisation
Obstructive sleep apnoea (OSA)	Characterised by a complete or partial collapse of the upper airways, leading to sleep fragmentation and intermittent hypoxemia.
Insomnia	Difficulty initiating or maintaining sleep despite having an opportunity for sleep. Classified as chronic when there are ≥ 3 nights of insomnia per week for at least 3 months.
REM sleep behaviour disorder (RBD)	Characterised by the loss of muscle atonia during REM sleep, resulting in abnormal motor manifestations or dream-enactment behaviour. It is a prodromal feature of Parkinson's disease and Dementia with Lewy Bodies, and multiple system atrophy. It is less common in AD and VaD and will not be addressed in this chapter.

downscaling of neuronal synapses. Synaptic downscaling enables further potentiation of synapses during wake, prevents oversaturation of synaptic strength, improves signal-to-noise ratio in neural circuits and conserves energy (14). In sleep deprivation studies, impaired synaptic downscaling leads to dysfunction in hippocampal synapses, and consequently, memory impairment (15). While REM sleep promotes adult hippocampal neurogenesis through boosting the processes of cell proliferation and cell survival (16). These processes are imperative for integrating new neuron cells into existing hippocampal circuitry to support memory (17). As described in detail in Chapter 3 within this book, sleep plays a significant role in the overnight consolidation of memories. It is, however, reduced in ageing (18) and maybe further compromised in MCI (19, 20) and dementia (21).

Various organs and tissues in the human body are synchronised to the circadian rhythm. Circadian rhythms may even regulate the blood-brain barrier, inflammation and oxidative stress (see review (22)). Mouse models show that the disruption of clock genes (e.g., circadian locomotor output cycles kaput (CLOCK) or brain and muscle ARNT-like protein 1 (BMAL)) is associated with blood-brain barrier hyper-permeability (23). They also indicate that circadian disruption leads to increased neuroinflammation (24). Antioxidant enzymes follow a circadian rhythm, suggesting that time of day may mediate oxidative stress (25). In turn, hyper-permeability of the blood-brain barrier (26), prolonged immune activation (27) and increased oxidative stress (28) are associated with AD pathology.

The more recent discovery of the 'glymphatic system' highlights another purpose for sleep. The glymphatic system is a glial-dependent pathway in the brain that appears to be integral to the clearance of the brain's waste products, including Aβ and tau proteins (29). This system enables the movement of CSF along perivascular spaces where it exchanges waste products with interstitial fluid within the brain parenchyma before being drained or 'flushed away' into the peripheral lymphatic system. Animal studies have shown that slow wave sleep enhances the exchange of waste products via the expansion and contraction of brain extracellular space by approximately 60%, compared to wakefulness (30). While further work in humans is required in order to better understand this system, it is likely that key neurometabolites and toxins produced during wakefulness are expelled during sleep.

In summary, there are postulated many mechanisms by which sleep disturbance may contribute to brain health and cognitive decline, including synaptic downscaling, adult neurogenesis, overnight memory consolidation, regulation of the immune system, oxidative stress and clearing of neurotoxins. Coupled with changes to sleep with ageing, it is possible that sleep disturbances play a direct causative role in neurodegeneration, leading to cognitive decline and dementia. It is also plausible that neurodegeneration itself disrupts key sleep circuits. The bidirectional relationships are yet to be elucidated, but literature to date linking sleep and neurodegeneration will be summarised.

Changes to sleep in ageing, MCI and dementia

There are various physiological changes that occur with increasing age, including shifts in sleep and circadian rhythms. The most common sleep-related changes reported by older adults include:

- Decreased total sleep time (31)
- Decreased deep sleep (slow wave sleep) (31)
- Decreased REM sleep (31)
- Increased number of awakenings after sleep onset (31)

- Decreased sleep efficiency (percentage of time spent asleep relative to time spent in bed) (31)
- Circadian phase advance (shifts to earlier sleep and wake times) (32)

Large epidemiological studies have shown that between 40 and 70% of older adults experience sleep disturbance or report having issues with their sleep (33). As defined in Table 4.1, as well as general sleep disturbance, sleep disorders are also highly prevalent in older adults, including insomnia, obstructive sleep apnoea (OSA), and REM sleep behaviour disorder (RBD). In older adults, up to 48% have insomnia (34), up to 51% have OSA (35) and 2% may meet criteria for RBD (36).

In people with MCI, subjective sleep complaints are even more pronounced than those seen in cognitively intact samples. Meta-analytic studies (37) of PSG changes have shown that relative to normal controls, those with MCI have:

- Decreased total sleep time
- Decreased sleep efficiency
- Increased time taken to fall asleep
- Decreased REM sleep
- Greater severity of nocturnal hypoxemia

In addition, studies using gold-standard measures of melatonin have shown circadian advancement in clinical samples with MCI (38), and using actigraphy, greater circadian rhythm fragmentation (39) has been found.

In older adults with dementia, key changes include:

- Increase sleep during day (40)
- Increase wake after sleep onset (40)
- Delayed circadian phase (40)
- Circadian changes such as sundowning, agitation and pineal changes (41)

These sleep changes can exacerbate cognitive impairment and potentially contribute to cognitive decline, behavioural disturbances and potentially to neurodegeneration.

Evidence linking sleep disturbances with cognitive decline, brain changes and potential underlying mechanisms

While there is a general paucity of longitudinal studies and other work affirming causality, there is an abundance of work linking poor sleep to cognitive function and cognitive decline. Here, we summarise findings in general poor sleep quality, efficiency and circadian alterations. See Figure 5.1 for proposed mechanisms linking sleep disturbances and disorders with AD and VaD.

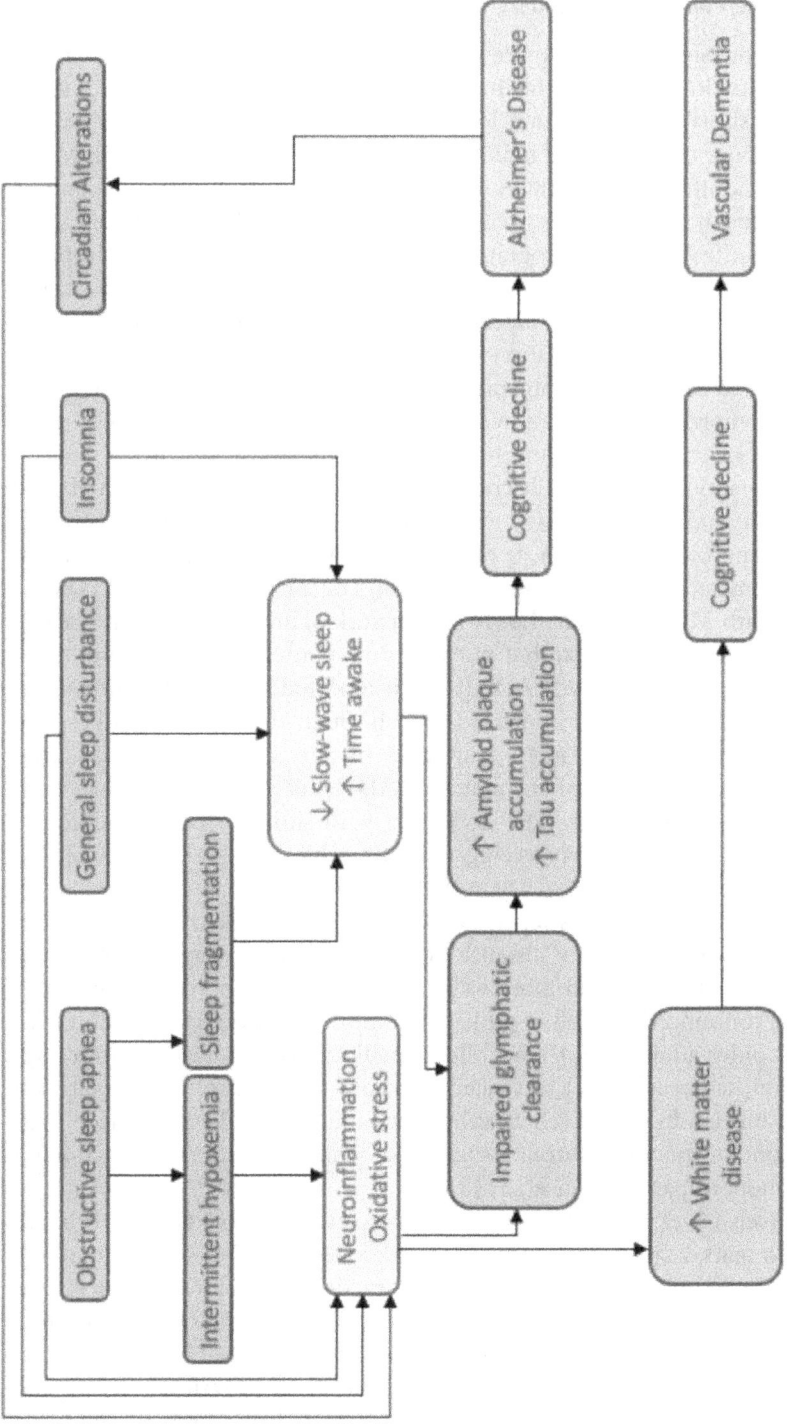

FIGURE 5.1   Proposed mechanisms linking various sleep disturbances and disorders with Alzheimer's Disease and Vascular Dementia. REM = Rapid Eye Movement.

General sleep disturbance, quality, efficiency

The National Sleep Foundation recommends 7–8 hours of sleep per night, but many adults do not follow this recommendation (42). Recent studies show that both short (<7 hours) and long (>8 hours) sleep durations are associated with poor cognition and brain health. For instance, a meta-analysis found that older adults were 1.86 times more likely to develop all-cause dementia if they had insufficient or excessive sleep duration (43). Winer and colleagues (44) demonstrated in 4417 cognitively intact older adults that short sleep was associated with greater Aβ burden, while long sleep was associated with executive dysfunction. Besides an increase in Aβ burden, sleep deprivation studies suggest that short sleep duration may cause the release of proinflammatory cytokines and influence synaptic pruning (45), which in turn have been linked to brain compromise and cognitive decline. It is possible that long sleep duration reflects a compensatory mechanism for poor sleep quality (46).

Sleep quality plays an equally critical role in brain health. Self-reported poor sleep quality has been associated with impaired executive function in community-dwelling older men (47). It has also been linked with the decoupling of the temporal and parietal regions of the brain's default mode network (DMN) in those with MCI (48). The DMN is a network of brain regions that is active when an individual is awake but not performing attention demanding tasks. The DMN is affected in early AD (49). Self-reported sleep disturbance is also associated with Aβ and tau burden (50), although the relationship between sleep and these AD biomarkers may differ by age. One recent study using the European Prevention of Dementia (EPAD) dataset (n=1240) showed that greater p-tau/Aβ42 ratios were associated with shorter sleep duration and higher sleep efficiency in middle-aged adults, while AD biomarker build-up was linked to longer sleep latency in the elderly (51).

Objective measures of sleep disturbance have supported the findings of these self-report studies. For instance, actigraphy-derived sleep fragmentation, poorer sleep efficiency and more wake after sleep onset have been linked to a more pronounced decline in global cognition and executive function (47) in 2822 older adults at 3.4-year follow-up, Furthermore, greater sleep fragmentation has been linked to thalamic atrophy and ventromedial prefrontal amyloid burden in cognitively healthy older adults (52). In MCI, sleep fragmentation and nocturnal arousals have been linked to executive functioning and memory impairment (n = 15) (53).

Relatively fewer studies have utilised PSG to examine how sleep disturbance relates to markers of brain health. Baril and colleagues (54) showed that in a sample of 492 cognitively intact older adults, decreased slow wave sleep was associated with smaller cortical and subcortical brain volumes and more white matter hyperintensities. In addition, Lucey and colleagues (55) showed that less NREM slow-wave activity was associated with tau (measured by PET and CSF), indicative of increased AD pathology in 119 older adults.

A common theme in these studies is that decreased sleep duration, sleep quality and altered sleep microarchitecture (slow-wave activity) are linked to AD pathology. However, data supporting causation is still lacking. There are few longitudinal studies in the field. However, there are some experimental studies that give credence to the notion that sleep disruption contributes to AD pathology. For instance, Ju and colleagues (56) demonstrated in 17 healthy adults that auditory disruptions to slow wave sleep caused an increase in CSF Aβ levels. However, animal studies also support the notion that neurodegenerative damage causes sleep disturbance. For example, in a mouse model, the spread of Aβ accumulation in the hippocampus and the cortex led to delayed sleep onset and increased wake after sleep onset (57). Overall, whilst evidence supporting a bidirectional relationship is mounting, further empirical longitudinal and experimental work is still required, and robust clinical trial evidence would also be informative.

Circadian alterations

The human circadian rhythm is regulated by several key brain regions, including the hypothalamus, thalamus, basal forebrain, brainstem and the cortex (58). Within the hypothalamus, the suprachiasmatic nucleus (the master circadian clock) follows an approximate 24-hour rhythm of melatonin secretion. In ageing, MCI and AD samples, disruption to this circadian rhythm has been linked to cognitive impairment and decline and to AD biomarkers. For example, in 189 older adults with preclinical AD (confirmed by PET imaging or CSF tau to Aβ42 ratio), sleep-wake rhythm fragmentation was associated with an increase in the CSF tau to Aβ42 ratio (59). In the large Study of Osteoporotic Fractures cohort, 1282 older women with actigraphy-defined delayed activity peak were 1.83 times more likely to develop dementia or MCI (60). Using dim light melatonin onset, people with MCI had circadian advance, relative to controls, which was in turn linked to poorer memory performance (38). In AD, fragmentation of the actigraphic rest-activity rhythm was associated with greater cognitive decline after one-year follow-up (61). This work suggests that altered circadian rhythms may precede dementia and are linked to cognitive decline longitudinally. More longitudinal studies are now required to determine whether circadian alterations merely coexist with neurodegeneration or play a causative role in disease pathogenesis.

The contribution of sleep disorders to cognitive decline, brain changes and potential underlying mechanisms

Obstructive sleep apnoea

There is accumulating evidence linking OSA to both cognitive decline and to brain alterations and intermittent hypoxemia and sleep fragmentation may play key mechanistic roles in brain compromise) (62). In middle-aged adults, OSA has been associated with various neuropsychological deficits, including in the domains of vigilance, attention, processing speed, episodic memory, working memory and executive function (see review (63)). Meta-analytic studies suggest the most striking relationships may be with memory with medium effect size (g=0.50) decrements found (64). Accordingly, changes in the hippocampus and para-hippocampal gyrus have also been found in meta-analyses (65).

The strength of the link between OSA and cognitive function in older adults may depend on the sampling frame. A meta-analysis by Cross and colleagues (66) showed only a small decrement between OSA and cognition overall, with the most pronounced (albeit small effect; g=0.18) relationships being evident in the memory domain. Of significance, however, moderate effect size (g=0.49) decrements were found when the analyses were restricted to clinical, as compared to community samples. However, even within community samples, OSA appears to predict later cognitive decline and dementia. Data from the Alzheimer's Disease Neuroimaging Initiative showed that OSA in older adults without previous cognitive impairment is linked to younger age of onset for both MCI and AD (67). An overall meta-analysis of 4288419 people and 14 studies showed that OSA in midlife is linked with a 26% increased risk of dementia (63).

The links between OSA and neuropsychological dysfunction may partially be explained by the extent of brain changes. For instance, in a memory clinic sample of 83 older adults with MCI and SCI, hypoxemia was associated with reduced cortical thickness in the bilateral temporal lobes, which in turn, mediated alterations in memory (68). Increases in brain volumes have also been found including in the postcentral gyrus, pericalcarine, pars opercularis, hippocampus, amygdala, entorhinal cortex, prefrontal cortex, parietal lobes, posterior cingulate and frontal gyrus (68, 69). Such increases have been postulated to reflect neuroinflammation and oedema. Indeed, links between OSA and increased entorhinal and hippocampal volumes dissipate when accounting for free water (unrestricted water molecules), which are linked to oedema and inflammation.

Abnormal functional connectivity in important brain networks has also been linked to OSA in older samples. In particular, para-hippocampal connectivity within the DMN has been associated with nocturnal intermittent hypoxemia and OSA in older adults with and without MCI (70). The DMN is involved in episodic memory, introspection and autobiographic memory (71). It is plausible that dysfunction within these critical brain networks may underpin the cognitive decline seen in OSA, though this has yet to be demonstrated.

Studies conducted in middle-aged and older adults show that OSA is also linked to various AD biomarkers, including those derived from PET, CSF and blood. In middle-aged and older adults, associations between OSA and increased Aβ and tau have been found (measured by both serum and PET imaging (47, 72)), as well as a decrease in CSF neuronal-derived proteins. In people with (n=798) and without (n=516) MCI, OSA was also associated with a faster AD biomarker accumulation over approximately 2.5 years (73).

While the cascade of events linking OSA to dementia is likely to be multifactorial, intermittent hypoxemia has been shown to increase Aβ generation (74) and increase levels of tau protein and tau hyperphosphorylation (75). Sleep fragmentation can also disrupt slow wave sleep, which is critical for the clearance of neurotoxins such as Aβ (76). Both have been linked to an increase in oxidative stress, microglial activation and neuroinflammation, all of which contribute to AD and vascular dementia (77). OSA may indirectly contribute to neurodegeneration by increasing the prevalence of other comorbidities, such as hypertension, and cardiometabolic syndrome (26). It can cause endothelial dysfunction, in turn, causing concomitant white matter damage and cerebrovascular disease (78). The link between OSA and cerebrovascular disease is particularly interesting and yet remains under-explored, particularly regarding how OSA may impact brain glymphatics. Hypertension, in particular, is exacerbated in patients with OSA because there is an increase in cerebral blood vessel stiffness (79), blood pressure surges that occur during apnoeas (80) and non-dipping blood pressure during sleep (81). Arterial stiffening causes non-compliance of blood vessels, which can cause an increase in blood pressure. The increase in microvascular pressure has been linked to impairment of glymphatic clearance (82). Taken together, the exacerbation of hypertension is a plausible mechanism by which OSA may contribute to dementia. Finally, it is important to note that OSA-related fragmentation of slow wave and REM sleep can disrupt overnight memory consolidation (20), therefore contributing to both short- and longer-term difficulties with encoding and storage of new memories.

Insomnia

Insomnia may also be linked to an increased risk of dementia. A meta-analysis of 48 studies found that adults with insomnia had impaired episodic memory, attention and executive function compared to controls, cross-sectionally (83). This relationship was the strongest in older samples. A meta-analysis of 51 studies showed that insomnia is associated with a 1.27 higher risk of developing cognitive disorders (84).

There is currently a lack of studies in preclinical, SCI or MCI samples examining whether insomnia is predictive of accelerated cognitive decline. Similarly, few studies have directly investigated the links between insomnia and neuroimaging markers of brain changes. One study conducted in a sample of 17 middle-aged insomnia subjects revealed decreased hippocampal volume (85) relative to 19 controls, but another (n=66) did not indicate differences in grey nor white matter volumes (86). One study of 54 participants showed increased functional connectivity between the left anterior cingulate cortex and the right thalamus (87).

There is some work examining the link between insomnia and neurodegeneration biomarkers. In a study of 46 middle-aged and older adults, compared to those without chronic insomnia, those with chronic insomnia had increased levels of Aβ42, but not Aβ40, total tau, or phosphorylated tau (88). Furthermore, higher serum neurofilament light concentration was elevated in middle-aged adults with chronic insomnia compared to controls (89), which is supportive of a general neurodegeneration process. Interestingly, AD pathology in older adults with comorbid insomnia has been linked to faster cognitive decline than AD pathology or insomnia alone (90), suggesting that having both AD and insomnia is associated with poorer outcomes.

Potential treatments for sleep and dementia and cognitive decline

Of key relevance for the scientific field and for clinical practice is to ascertain how and when sleep disturbances can be best detected and treated in at-risk older samples in order to optimise cognition and potentially even delay or slow dementia risk.

Interventions for obstructive sleep apnoea

There are several interventions for the management of OSA, the most common being continuous positive airway pressure (CPAP), mandibular advancement splints (MAS) and weight loss. CPAP delivers constant air pressure to maintain open airways during sleep. MAS is an oral appliance that moves the lower jaw slightly forward to prevent upper airway obstruction during sleep. These therapies are effective in reducing OSA severity, if treatment adherence is adequate (91).

In middle-aged samples, CPAP findings are mixed, with studies indicating null (92) or positive effects on memory or executive functioning (93, 94). One study indicated positive benefits of MAS on processing speed (95). In ageing or MCI samples, studies are emerging and underway (e.g., see ANZCTR 12621001190897) in this field. Recently, Hoyos and colleagues (2022) showed that in a 12-week pilot randomised crossover clinical trial, CPAP did not improve executive function or processing speed in MCI, but small improvements were observed in memory (96). In the MEMORIES study, Richards and colleagues (2019) showed that in 54 people with MCI, those who were adherent to CPAP (at least four hours a night) for one-year, improvements in processing speed, memory and attention were evident compared to the non-adherent group (97). However, findings are less clear when examining incident rates of dementia. In a retrospective study of MCI and OSA (n=96), Skiba and colleagues reported CPAP use did not delay progression to dementia at approximately 2.8 years follow-up (98). However, in a larger, retrospective study of 53,321 older adults with OSA, those who had CPAP treatment and were adherent, had a lower incidence of AD diagnoses (99).

Cognitive behavioural therapy for insomnia

Cognitive behaviour therapy for insomnia (CBT-I) aims to adjust maladaptive thinking and behaviours that contribute to insomnia) and is the first-line treatment for chronic insomnia. In adult studies (mean age=56 years old), CBT-I shows large effect size improvements in insomnia symptoms, reducing wake after sleep onset and improving sleep quality (see review (100)).

While there are not yet large robust studies of CBT-I in older at-risk samples, emerging evidence from three small studies supports the use of CBT-I for sleep in people with MCI but benefits for cognition are unclear. In one study (101) of 35 people with MCI and sleep disturbances, group-based CBT-I demonstrated large to moderate effect size improvements in subjective sleep quality and daytime sleepiness, compared to control, and small effect sizes were noted for executive functions (not memory). By contrast, in 28 older adults with comorbid MCI and insomnia, six sessions of CBT-I were linked with improved sleep and executive functioning (102). Further trials are underway investigating the utility of digital CBT-I programs (see NCT05568381, NCT05565833 and NCT05173844).

Physical and cognitive activity to improve sleep quality

Physical activity has been shown to improve sleep quality (see review (103)) in both young and older adults. Only one known randomised-controlled trial (n=60) specifically targeting sleep quality has been conducted in ageing/MCI. The results of the 20-week trial showed that exercise was associated with improvements in both sleep quality and global cognition relative to a control group; however, it is unclear whether there was a relationship between the improvement in sleep quality and cognition (104).

There is some limited work suggesting that cognitive interventions may improve sleep, as well as cognition. One study showed a cognitive intervention program improved memory, mood and subjective sleep quality (105) and others have linked cognitive interventions to reduced daytime sleepiness (106) and improved sleep quality (106).

Neuromodulation to boost sleep microarchitecture

Neuromodulation has been shown to boost slow sleep oscillations and overnight memory consolidation. Common techniques used in neuromodulation include transcranial electrical stimulation (TES), transcranial magnetic stimulation (TMS) and acoustic stimulation.

There are several forms of TES that involve the application of mild electrical currents onto the scalp. The most common form is transcranial direction current stimulation (tDCS), which utilises a constant current. In younger and older samples, tDCS has been used to boost slow oscillations, sleep spindles and temporal coupling of slow oscillations and sleep spindles. In older adults, several small studies (n=~20) have indicated an improvement in overnight memory consolidation with tDCS (107, 108) and a similar finding was reported in one small study (n=16) of amnestic MCI (109). Transcranial alternating current stimulation (tACS) applies alternating currents to the scalp, and in younger adults, it appears to improve slow oscillations and overnight memory consolidation (110). No known studies have examined tACS in older adults. TMS utilises a magnetic field to send currents to the cerebral cortex to either increase or decrease neuronal excitability. Although difficult to conduct during sleep, Massimini and colleagues (2007) (111) demonstrated that TMS can increase slow-wave activity during sleep in 15 young males. Studies examining older adults and effects on cognition are yet to be conducted.

Acoustic stimulation utilises the brain's receptiveness to auditory cues during sleep to enhance and increase the number of slow oscillations and spindle activity during sleep. A small study of 13 older adults demonstrated that acoustic stimulation increased slow oscillation and spindle density, as well as improved declarative overnight memory consolidation (112). In another small study of nine older adults with amnestic MCI, acoustic stimulation caused an increase in slow-wave activity, which was related to better declarative overnight memory consolidation (113).

Other pharmacological sleep treatments

Extensive reviews of pharmacological therapies for sleep are available (see review (114)). Only those believed to be most relevant to cognition and dementia will be examined, including the benzodiazepines, Z-drugs orexin antagonists and melatonin.

Benzodiazepines bind to GABA-$_A$ receptors to sedate and reduce sleep latency and wake after sleep onset. However, the side effects of benzodiazepines can be quite severe. They suppress N3 and REM sleep, which are the most restorative stages of sleep. Furthermore, use of benzodiazepines in older adults has been linked to falls, confusion and greater risk of developing dementia (see meta-analysis by He et al., 2019 (115)). Subsequently, the use of benzodiazepines in older adults is not recommended.

Some work suggests that the group of benzodiazepine-like drugs (e.g., zolpidem, eszopiclone and zopiclone) termed 'Z-drugs' can boost sleep spindles and, in turn, improve overnight memory consolidation in young adults (116). Like benzodiazepines, Z-drugs enhance the GABA neurotransmitter (although binds to a different GABA-A receptor site). The half-life of Z-drugs is shorter than benzodiazepines. Thus, the side effects are less severe. There is inconsistent literature concerning Z-drug use and AD risk. One large study (n=353,581) showed that the use of Z-drugs increased AD risk after five years (117), whilst another (n=2,876) found a non-significant link (118). Thus, further trials are now needed to determine the brain and cognitive effects of Z-drug use in older people and those at-risk of dementia.

Orexin antagonists block one or both orexin receptors to promote sleep. Dual orexin receptor antagonists are effective in treating primary insomnia, decreasing wake after sleep onset, increasing sleep efficiency and total sleep time (see meta-analysis by Xue et al. (119)). Orexin antagonists have been shown to decrease Aâ plaque formation and improve cognition (120). While preliminary evidence for orexin antagonists is promising and side effects are minimal (121), there is at this stage no evidence that this drug improves cognition or lowers risk for cognitive decline or dementia.

Melatonin supplements increase the availability of the melatonin hormone that is naturally produced by the pineal gland. It can reduce sleep latency, increase total sleep time and improve sleep quality (see (122) for a meta-analysis). In dementia, melatonin improved sleep efficiency, and increased total sleep time, albeit with null effects on cognitive function (see (123) for a review). In MCI samples, some work suggests that melatonin may improve memory (124), executive function (124), processing speed (124), mood (124), sleep quality (124, 125) and circadian rhythms (125). Furthermore, melatonin may have anti-inflammatory and antioxidant properties and has been linked to reduced inflammation as measured by interleukin-1, interleukin-6, interluekin-8 and tumour necrosis factor alpha (see (126) for meta-analysis) and a reduction in oxidative stress parameters including decreased lipid peroxidation and increased total antioxidant capacity (see (127) for a meta-analysis). However, to date, no studies have examined whether the improved sleep and anti-inflammatory and antioxidant properties have positive benefits for reducing dementia risk.

Conclusion

Sleep contributes to maintaining healthy brain function. Various sleep disturbances and disorders may disrupt the benefits of sleep for brain health and cognition, possibly causing a cascade of events that contribute to dementia. Of all the sleep disturbances and disorders, OSA is the most well-examined with regard to cognition, brain changes and dementia risk. More studies are necessary for more generalised sleep disturbance (quality, duration efficiency), as well as for insomnia and circadian misalignment. Only preliminary evidence is available for the efficacy of sleep interventions in older at-risk people, and there is a dearth of data available to show there are subsequent benefits for cognition and dementia risk. There is a clear need for definitive clinical trials to confirm these findings.

References

1 Global mortality from dementia: application of a new method and results from the Global Burden of Disease Study 2019. *Alzheimers Dement.* 2021;7(1):e12200. https://doi.org/10.1002/trc2.12200
2 Barker WW, Luis CA, Kashuba A, et al. Relative frequencies of Alzheimer disease, Lewy body, vascular and frontotemporal dementia, and hippocampal sclerosis in the State of Florida Brain Bank. *Alzheimer Dis Assoc Disord.* 2002;16(4):203–12. https://doi.org/10.1097/00002093-200210000-00001
3 Goodman RA, Lochner KA, Thambisetty M, Wingo TS, Posner SF, Ling SM. Prevalence of dementia subtypes in United States Medicare fee-for-service beneficiaries, 2011–2013. *Alzheimers Dement.* 2017;13(1):28–37. https://doi.org/10.1016/j.jalz.2016.04.002
4 Brunnström H, Gustafson L, Passant U, Englund E. Prevalence of dementia subtypes: a 30-year retrospective survey of neuropathological reports. *Arch Gerontol Geriatr.* 2009;49(1):146–9. https://doi.org/10.1016/j.archger.2008.06.005
5 Attems J, Jellinger KA. The overlap between vascular disease and Alzheimer's disease – lessons from pathology. *BMC Med.* 2014;12(1):206. https://doi.org/10.1186/s12916-014-0206-2
6 Lippens G, Sillen A, Landrieu I, et al. Tau aggregation in Alzheimer's disease: what role for phosphorylation? *Prion.* 2007;1(1):21–5. https://doi.org/10.4161/pri.1.1.4055
7 Reisberg B, Shulman MB, Torossian C, Leng L, Zhu W. Outcome over seven years of healthy adults with and without subjective cognitive impairment. *Alzheimers Dement.* 2010;6(1):11–24. https://doi.org/10.1016/j.jalz.2009.10.002
8 Long JM, Coble DW, Xiong C, et al. Preclinical Alzheimer's disease biomarkers accurately predict cognitive and neuropathological outcomes. *Brain.* 2022. https://doi.org/10.1093/brain/awac250
9 Knopman DS, Jack CR, Jr., Wiste HJ, et al. Short-term clinical outcomes for stages of NIA-AA preclinical Alzheimer disease. *Neurology.* 2012;78(20):1576–82. https://doi.org/10.1212/WNL.0b013e3182563bbe
10 Mitchell AJ, Shiri-Feshki M. Rate of progression of mild cognitive impairment to dementia – meta-analysis of 41 robust inception cohort studies. *Acta Psychiatr Scand.* 2009;119(4):252–65. https://doi.org/10.1111/j.1600-0447.2008.01326.x
11 van Dyck CH, Swanson CJ, Aisen P, et al. Lecanemab in early Alzheimer's disease. *N Engl J Med.* 2022;388(1):9–21. https://doi.org/10.1056/NEJMoa2212948
12 Vermunt L, Sikkes SAM, van den Hout A, et al. Duration of preclinical, prodromal, and dementia stages of Alzheimer's disease in relation to age, sex, and APOE genotype. *Alzheimers Dement.* 2019;15(7):888–98. https://doi.org/10.1016/j.jalz.2019.04.001
13 Livingston G, Huntley J, Sommerlad A, et al. Dementia prevention, intervention, and care: 2020 report of the Lancet commission. *Lancet.* 2020;396(10248):413–46. https://doi.org/10.1016/S0140-6736(20)30367-6
14 Tononi G, Cirelli C. Sleep and synaptic homeostasis: a hypothesis. *Brain Res Bull.* 2003;62(2):143–50. https://doi.org/10.1016/j.brainresbull.2003.09.004
15 Cirelli C. Sleep and synaptic changes. *Curr Opin Neurobiol.* 2013;23(5):841–6. https://doi.org/10.1016/j.conb.2013.04.001
16 Navarro-Sanchis C, Brock O, Winsky-Sommerer R, Thuret S. Modulation of adult hippocampal neurogenesis by sleep: impact on mental health review. *Front Neural Circuits.* 2017;11. https://doi.org/10.3389/fncir.2017.00074
17 Toni N, Teng EM, Bushong EA, et al. Synapse formation on neurons born in the adult hippocampus. *Nat Neurosci.* 2007;10(6):727–34. https://doi.org/10.1038/nn1908

18 Gui WJ, Li HJ, Guo YH, Peng P, Lei X, Yu J. Age-related differences in sleep-based memory consolidation: a meta-analysis. *Neuropsychologia.* 2017;97:46–55. https://doi.org/10.1016/j.neuropsychologia.2017.02.001

19 Westerberg CE, Mander BA, Florczak SM, et al. Concurrent impairments in sleep and memory in amnestic mild cognitive impairment. *J Int Neuropsychol Soc.* 2012;18(3):490–500. https://doi.org/10.1017/s135561771200001x

20 Lam A, Haroutonian C, Grummitt L, et al. Sleep-dependent memory in older people with and without MCI: the relevance of sleep microarchitecture, OSA, hippocampal subfields, and episodic memory. *Cereb Cortex.* 2021;31(6):2993–3005. https://doi.org/10.1093/cercor/bhaa406

21 Rauchs G, Schabus M, Parapatics S, et al. Is there a link between sleep changes and memory in Alzheimer's disease? *Neuroreport.* 2008;19(11):1159–62. https://doi.org/10.1097/WNR.0b013e32830867c4

22 Leng Y, Musiek ES, Hu K, Cappuccio FP, Yaffe K. Association between circadian rhythms and neurodegenerative diseases. *Lancet Neurol.* 2019;18(3):307–18. https://doi.org/10.1016/s1474-4422(18)30461-7

23 Nakazato R, Kawabe K, Yamada D, et al. Disruption of Bmal1 impairs blood-brain barrier integrity via pericyte dysfunction. *J Neurosci.* 2017;37(42):10052–62. https://doi.org/10.1523/jneurosci.3639-16.2017

24 Lauretti E, Di Meco A, Merali S, Praticò D. Circadian rhythm dysfunction: a novel environmental risk factor for Parkinson's disease. *Mol Psychiatry.* 2017;22(2):280–6. https://doi.org/10.1038/mp.2016.47

25 Edgar RS, Green EW, Zhao Y, et al. Peroxiredoxins are conserved markers of circadian rhythms. *Nature.* 2012;485(7399):459–64. https://doi.org/10.1038/nature11088

26 Al Lawati NM, Patel SR, Ayas NT. Epidemiology, risk factors, and consequences of obstructive sleep apnea and short sleep duration. *Prog Cardiovasc Dis.* 2009;51(4):285–93. https://doi.org/10.1016/j.pcad.2008.08.001

27 Kinney JW, Bemiller SM, Murtishaw AS, Leisgang AM, Salazar AM, Lamb BT. Inflammation as a central mechanism in Alzheimer's disease. *Alzheimers Dement.* 2018;4:575–90. https://doi.org/10.1016/j.trci.2018.06.014

28 Tönnies E, Trushina E. Oxidative stress, synaptic dysfunction, and Alzheimer's disease. *J Alzheimers Dis.* 2017;57(4):1105–21. https://doi.org/10.3233/jad-161088

29 Jessen NA, Munk AS, Lundgaard I, Nedergaard M. The glymphatic system: a beginner's guide. *Neurochem Res.* 2015;40(12):2583–99. https://doi.org/10.1007/s11064-015-1581-6

30 Xie L, Kang H, Xu Q, et al. Sleep drives metabolite clearance from the adult brain. *Science.* 2013;342(6156):373–7. https://doi.org/10.1126/science.1241224

31 Li J, Vitiello MV, Gooneratne NS. Sleep in normal aging. *Sleep Med Clin.* 2018;13(1):1–11. https://doi.org/10.1016/j.jsmc.2017.09.001

32 Duffy JF, Zitting KM, Chinoy ED. Aging and circadian rhythms. *Sleep Med Clin.* 2015;10(4):423–34. https://doi.org/10.1016/j.jsmc.2015.08.002

33 Miner B, Kryger MH. Sleep in the aging population. *Sleep Med Clin.* 2017;12(1):31–8. https://doi.org/10.1016/j.jsmc.2016.10.008

34 Patel D, Steinberg J, Patel P. Insomnia in the elderly: a review. *J Clin Sleep Med.* 2018;14(6):1017–24. https://doi.org/10.5664/jcsm.7172

35 Young T, Shahar E, Nieto FJ, et al. Predictors of sleep-disordered breathing in community-dwelling adults: the sleep heart health study. *Arch Intern Med.* 2002;162(8):893–900. https://doi.org/10.1001/archinte.162.8.893

36 Kang SH, Yoon IY, Lee SD, Han JW, Kim TH, Kim KW. REM sleep behavior disorder in the Korean elderly population: prevalence and clinical characteristics. *Sleep.* 2013;36(8):1147–52. https://doi.org/10.5665/sleep.2874

37 D'Rozario AL, Chapman JL, Phillips CL, et al. Objective measurement of sleep in mild cognitive impairment: a systematic review and meta-analysis. *Sleep Med Rev.* 2020;52:101308. https://doi.org/10.1016/j.smrv.2020.101308

38 Naismith SL, Hickie IB, Terpening Z, et al. Circadian misalignment and sleep disruption in mild cognitive impairment. *J Alzheimers Dis.* 2014;38(4):857–66. https://doi.org/10.3233/jad-131217

39 Musiek ES, Bhimasani M, Zangrilli MA, Morris JC, Holtzman DM, Ju YS. Circadian rest-activity pattern changes in aging and preclinical Alzheimer disease. *JAMA Neurol.* 2018;75(5):582–90. https://doi.org/10.1001/jamaneurol.2017.4719

40 Hatfield CF, Herbert J, van Someren EJ, Hodges JR, Hastings MH. Disrupted daily activity/rest cycles in relation to daily cortisol rhythms of home-dwelling patients with early Alzheimer's dementia. *Brain.* 2004;127(Pt 5):1061–74. https://doi.org/10.1093/brain/awh129

41 Volicer L, Harper DG, Manning BC, Goldstein R, Satlin A. Sundowning and circadian rhythms in Alzheimer's disease. *Am J Psychiatry.* 2001;158(5):704–11. https://doi.org/10.1176/appi.ajp.158.5.704

42 Hirshkowitz M, Whiton K, Albert SM, et al. National Sleep Foundation's updated sleep duration recommendations: final report. *Sleep Health.* 2015;1(4):233–43. https://doi.org/10.1016/j.sleh.2015.10.004

43 Bubu OM, Brannick M, Mortimer J, et al. Sleep, cognitive impairment, and Alzheimer's disease: a systematic review and meta-analysis. *Sleep.* 2017;40(1). https://doi.org/10.1093/sleep/zsw032

44 Winer JR, Deters KD, Kennedy G, et al. Association of short and long sleep duration with amyloid-β burden and cognition in aging. *JAMA Neurol.* 2021;78(10):1187–96. https://doi.org/10.1001/jamaneurol.2021.2876

45 Tuan L-H, Lee L-J. Microglia-mediated synaptic pruning is impaired in sleep-deprived adolescent mice. *Neurobiol Dis.* 2019;130:104517. https://doi.org/10.1016/j.nbd.2019.104517

46 Bin YS. Is sleep quality more important than sleep duration for public health? *Sleep.* 2016;39(9):1629–30. https://doi.org/10.5665/sleep.6078

47 Blackwell T, Yaffe K, Laffan A, et al. Associations of objectively and subjectively measured sleep quality with subsequent cognitive decline in older community-dwelling men: the MrOS sleep study. *Sleep.* 2014;37(4):655–63. https://doi.org/10.5665/sleep.3562

48 McKinnon AC, Lagopoulos J, Terpening Z, et al. Sleep disturbance in mild cognitive impairment is associated with alterations in the brain's default mode network. *Behav Neurosci.* 2016;130(3):305–15. https://doi.org/10.1037/bne0000137

49 Palmqvist S, Schöll M, Strandberg O, et al. Earliest accumulation of β-amyloid occurs within the default-mode network and concurrently affects brain connectivity. *Nat Commun.* 2017;8(1):1214. https://doi.org/10.1038/s41467-017-01150-x

50 Winer JR, Mander BA, Helfrich RF, et al. Sleep as a potential biomarker of tau and β-amyloid burden in the human brain. *J Neurol Sci.* 2019;39(32):6315–24. https://doi.org/10.1523/jneurosci.0503-19.2019

51 Naismith SL, Leng Y, Palmer JR, Lucey BP. Age differences in the association between sleep and Alzheimer's disease biomarkers in the EPAD cohort. *Alzheimers Dement: DADM.* 2022;14(1):e12380. https://doi.org/10.1002/dad2.12380

52 André C, Tomadesso C, de Flores R, et al. Brain and cognitive correlates of sleep fragmentation in elderly subjects with and without cognitive deficits. *Alzheimers Dement.* 2019;11:142–50. https://doi.org/10.1016/j.dadm.2018.12.009

53 Naismith SL, Rogers NL, Hickie IB, Mackenzie J, Norrie LM, Lewis SJ. Sleep well, think well: sleep-wake disturbance in mild cognitive impairment. *J Geriatr Psychiatry Neurol.* 2010;23(2):123–30. https://doi.org/10.1177/0891988710363710

54 Baril AA, Beiser AS, Mysliwiec V, et al. Slow-wave sleep and MRI markers of brain aging in a community-based sample. *Neurology.* 2021;96(10):e1462–9. https://doi.org/10.1212/wnl.0000000000011377

55 Lucey BP, McCullough A, Landsness EC, et al. Reduced non-rapid eye movement sleep is associated with tau pathology in early Alzheimer's disease. *Sci Transl Med.* 2019;11(474). https://doi.org/10.1126/scitranslmed.aau6550

56 Ju Y-ES, Ooms SJ, Sutphen C, et al. Slow wave sleep disruption increases cerebrospinal fluid amyloid-β levels. *Brain.* 2017;140(8):2104–11. https://doi.org/10.1093/brain/awx148

57 Roh JH, Huang Y, Bero AW, et al. Disruption of the sleep-wake cycle and diurnal fluctuation of β-amyloid in mice with Alzheimer's disease pathology. *Sci Transl Med.* 2012;4(150):150ra122. https://doi.org/10.1126/scitranslmed.3004291

58 Porkka-Heiskanen T, Zitting K-M, Wigren H-K. Sleep, its regulation and possible mechanisms of sleep disturbances. *Acta Physiol.* 2013;208(4):311–28. https://doi.org/10.1111/apha.12134

59 Musiek ES, Bhimasani M, Zangrilli MA, Morris JC, Holtzman DM, Ju Y-ES. Circadian rest-activity pattern changes in aging and preclinical Alzheimer disease. *JAMA Neurol.* 2018;75(5):582–90. https://doi.org/10.1001/jamaneurol.2017.4719

60 Tranah GJ, Blackwell T, Stone KL, et al. Circadian activity rhythms and risk of incident dementia and mild cognitive impairment in older women. *Ann Neurol.* 2011;70(5):722–32. https://doi.org/10.1002/ana.22468

61 Targa ADS, Benítez ID, Dakterzada F, et al. The circadian rest-activity pattern predicts cognitive decline among mild-moderate Alzheimer's disease patients. *Alzheimer's Res Ther.* 2021;13(1):161. https://doi.org/10.1186/s13195-021-00903-7

62 Naismith S, Winter V, Gotsopoulos H, Hickie I, Cistulli P. Neurobehavioral functioning in obstructive sleep apnea: differential effects of sleep quality, hypoxemia and subjective sleepiness. *J Clin Exp Neuropsychol.* 2004;26(1):43–54. https://doi.org/10.1076/jcen.26.1.43.23929

63 Leng Y, McEvoy CT, Allen IE, Yaffe K. Association of sleep-disordered breathing with cognitive function and risk of cognitive impairment: a systematic review and meta-analysis. *JAMA Neurol.* 2017;74(10):1237–45. https://doi.org/10.1001/jamaneurol.2017.2180

64 Wallace A, Bucks RS. Memory and obstructive sleep apnea: a meta-analysis. *Sleep.* 2013;36(2):203–20. https://doi.org/10.5665/sleep.2374

65 Huang X, Tang S, Lyu X, Yang C, Chen X. Structural and functional brain alterations in obstructive sleep apnea: a multimodal meta-analysis. *Sleep Med.* 2019;54:195–204. https://doi.org/10.1016/j.sleep.2018.09.025

66 Cross N, Lampit A, Pye J, Grunstein RR, Marshall N, Naismith SL. Is obstructive sleep apnoea related to neuropsychological function in healthy older adults? A systematic review and meta-analysis. *Neuropsychol Rev.* 2017;27(4):389–402. https://doi.org/10.1007/s11065-017-9344-6

67 Osorio RS, Gumb T, Pirraglia E, et al. Sleep-disordered breathing advances cognitive decline in the elderly. *Neurology.* 2015;84(19):1964–71. https://doi.org/10.1212/wnl.0000000000001566

68 Cross NE, Memarian N, Duffy SL, et al. Structural brain correlates of obstructive sleep apnoea in older adults at risk for dementia. *Eur Respir J.* 2018;52(1):1800740. https://doi.org/10.1183/13993003.00740-2018

69 Baril AA, Gagnon K, Brayet P, et al. Gray matter hypertrophy and thickening with obstructive sleep apnea in middle-aged and older adults. *Am J Respir Crit Care Med.* 2017;195(11):1509–18. https://doi.org/10.1164/rccm.201606-1271OC

70 Naismith SL, Duffy SL, Cross N, et al. Nocturnal hypoxemia is associated with altered parahippocampal functional brain connectivity in older adults at risk for dementia. *J Alzheimers Dis.* 2020;73(2):571–84. https://doi.org/10.3233/jad-190747

71 Li W, Mai X, Liu C. The default mode network and social understanding of others: what do brain connectivity studies tell us review. *Front Hum Neurosci.* 2014;8. https://doi.org/10.3389/fnhum.2014.00074

72 Jackson ML, Cavuoto M, Schembri R, et al. Severe Obstructive Sleep Apnea is associated with higher brain amyloid burden: a preliminary PET imaging study. *J Alzheimers Dis.* 2020;78(2):611–17. https://doi.org/10.3233/jad-200571

73 Bubu OM, Pirraglia E, Andrade AG, et al. Obstructive sleep apnea and longitudinal Alzheimer's disease biomarker changes. *Sleep.* 2019;42(6). https://doi.org/10.1093/sleep/zsz048

74 Salminen A, Kauppinen A, Kaarniranta K. Hypoxia/ischemia activate processing of Amyloid Precursor Protein: impact of vascular dysfunction in the pathogenesis of Alzheimer's disease. *J Neurochem.* 2017;140(4):536–49. https://doi.org/10.1111/jnc.13932

75 Zhang CE, Yang X, Li L, et al. Hypoxia-induced tau phosphorylation and memory deficit in rats. *Neurodegener Dis.* 2014;14(3):107–16. https://doi.org/10.1159/000362239

76 Ju Y-ES, Ooms SJ, Sutphen C, et al. Slow wave sleep disruption increases cerebrospinal fluid amyloid-β levels. *Brain.* 2017;140(8):2104–11. https://doi.org/10.1093/brain/awx148

77 Polsek D, Gildeh N, Cash D, et al. Obstructive sleep apnoea and Alzheimer's disease: in search of shared pathomechanisms. *Neurosci Biobehav Rev.* 2018;86:142–9. https://doi.org/10.1016/j.neubiorev.2017.12.004

78 Badji A, Noriega de la Colina A, Karakuzu A, et al. Arterial stiffness cut-off value and white matter integrity in the elderly. *Neuroimage Clin.* 2020;26:102007. https://doi.org/10.1016/j.nicl.2019.102007

79 Lee WJ, Jung KH, Nam HW, Lee YS. Effect of obstructive sleep apnea on cerebrovascular compliance and cerebral small vessel disease. *PLoS ONE.* 2021;16(11):e0259469. https://doi.org/10.1371/journal.pone.0259469

80 Marrone O, Bonsignore MR. Blood-pressure variability in patients with obstructive sleep apnea: current perspectives. *Nat Sci Sleep.* 2018;10:229–42. https://doi.org/10.2147/nss.S148543

81 Cuspidi C, Tadic M, Sala C, Gherbesi E, Grassi G, Mancia G. Blood pressure non-dipping and obstructive sleep apnea syndrome: a meta-analysis. *J Clin Med.* 2019;8(9):1367. https://doi.org/10.3390/jcm8091367

82 Ungvari Z, Toth P, Tarantini S, et al. Hypertension-induced cognitive impairment: from pathophysiology to public health. *Nat Rev Nephrol.* 2021;17(10):639–54. https://doi.org/10.1038/s41581-021-00430-6

83 Wardle-Pinkston S, Slavish DC, Taylor DJ. Insomnia and cognitive performance: a systematic review and meta-analysis. *Sleep Med Rev.* 2019;48:101205. https://doi.org/10.1016/j.smrv.2019.07.008

84 Xu W, Tan C-C, Zou J-J, Cao X-P, Tan L. Sleep problems and risk of all-cause cognitive decline or dementia: an updated systematic review and meta-analysis. *J Neurol Neurosurg Psychiatry.* 2020;91(3):236–44. https://doi.org/10.1136/jnnp-2019-321896

85 Neylan TC, Mueller SG, Wang Z, et al. Insomnia severity is associated with a decreased volume of the CA3/dentate gyrus hippocampal subfield. *Biol Psychiatry.* 2010;68(5):494–6. https://doi.org/10.1016/j.biopsych.2010.04.035

86 Spiegelhalder K, Regen W, Baglioni C, et al. Insomnia does not appear to be associated with substantial structural brain changes. *Sleep.* 2013;36(5):731–7. https://doi.org/10.5665/sleep.2638

87 Yan CQ, Liu CZ, Wang X, et al. Abnormal functional connectivity of anterior cingulate cortex in patients with primary insomnia: a resting-state functional magnetic resonance imaging study. *Front Aging Neurosci.* 2018;10:167. https://doi.org/10.3389/fnagi.2018.00167

88 Chen DW, Wang J, Zhang LL, Wang YJ, Gao CY. Cerebrospinal fluid amyloid-β levels are increased in patients with insomnia. *J Alzheimers Dis.* 2018;61(2):645–51. https://doi.org/10.3233/jad-170032

89 Insel PS, Mohlenhoff BS, Neylan TC, Krystal AD, Mackin RS. Association of sleep and β-amyloid pathology among older cognitively unimpaired adults. *JAMA Netw Open.* 2021;4(7):e2117573. https://doi.org/10.1001/jamanetworkopen.2021.17573

90 Pang R, Zhan Y, Zhang Y, et al. Aberrant functional connectivity architecture in participants with chronic insomnia disorder accompanying cognitive dysfunction: a whole-brain, data-driven analysis. *Front Neurosci.* 2017;11:259. https://doi.org/10.3389/fnins.2017.00259

91 Spicuzza L, Caruso D, Di Maria G. Obstructive sleep apnoea syndrome and its management. *Ther Adv Chronic Dis.* 2015;6(5):273–85. https://doi.org/10.1177/2040622315590318

92 Dostálová V, Kolečkárová S, Kuška M, Pretl M, Bezdicek O. Effects of continuous positive airway pressure on neurocognitive and neuropsychiatric function in obstructive sleep apnea. *J Sleep Res.* 2019;28(5):e12761. https://doi.org/10.1111/jsr.12761

93 Kushida CA, Nichols DA, Holmes TH, et al. Effects of continuous positive airway pressure on neurocognitive function in obstructive sleep apnea patients: the Apnea Positive Pressure Long-term Efficacy Study (APPLES). *Sleep.* 2012;35(12):1593–602. https://doi.org/10.5665/sleep.2226

94 Jackson ML, McEvoy RD, Banks S, Barnes M. Neurobehavioral impairment and CPAP treatment response in mild-moderate obstructive sleep apnea. *J Clin Sleep Med.* 2018;14(1):47–56. https://doi.org/10.5664/jcsm.6878

95 Naismith SL, Winter VR, Hickie IB, Cistulli PA. Effect of oral appliance therapy on neurobehavioral functioning in obstructive sleep apnea: a randomized controlled trial. *J Clin Sleep Med.* 2005;1(4):374–80.

96 Hoyos CM, Cross NE, Terpening Z, et al. Continuous positive airway pressure for cognition in sleep apnea and mild cognitive impairment: a pilot randomized crossover clinical trial. *Am J Respir Crit Care Med.* 2022;205(12):1479–82. https://doi.org/10.1164/rccm.202111-2646LE

97 Richards KC, Gooneratne N, Dicicco B, et al. CPAP adherence may slow 1-year cognitive decline in older adults with mild cognitive impairment and apnea. *J Am Geriatr Soc.* 2019;67(3):558–64. https://doi.org/10.1111/jgs.15758

98 Skiba V, Novikova M, Suneja A, McLellan B, Schultz L. Use of positive airway pressure in mild cognitive impairment to delay progression to dementia. *J Clin Sleep Med.* 2020;16(6):863–70. https://doi.org/10.5664/jcsm.8346

99 Dunietz GL, Chervin RD, Burke JF, Conceicao AS, Braley TJ. Obstructive sleep apnea treatment and dementia risk in older adults. *Sleep.* 2021;44(9). https://doi.org/10.1093/sleep/zsab076

100 Trauer JM, Qian MY, Doyle JS, Rajaratnam SM, Cunnington D. Cognitive behavioral therapy for chronic insomnia: a systematic review and meta-analysis. *Ann Intern Med.* 2015;163(3):191–204. https://doi.org/10.7326/m14-2841

101 Naismith SL, Pye J, Terpening Z, Lewis S, Bartlett D. "Sleep well, think well" group program for mild cognitive impairment: a randomized controlled pilot study. *Behav Sleep Med.* 2019;17(6):778–89. https://doi.org/10.1080/15402002.2018.1518223

102 Cassidy-Eagle E, Siebern A, Unti L, Glassman J, O'Hara R. Neuropsychological functioning in older adults with mild cognitive impairment and insomnia randomized to CBT-I or control group. *Clin Gerontol.* 2018;41(2):136–44. https://doi.org/10.1080/07317115.2017.1384777

103 Wang F, Boros S. The effect of physical activity on sleep quality: a systematic review. *Eur J Phys.* 2021;23(1):11–18. https://doi.org/10.1080/21679169.2019.1623314

104 Bademli K, Lok N, Canbaz M, Lok S. Effects of physical activity program on cognitive function and sleep quality in elderly with mild cognitive impairment: a randomized controlled trial. *Perspect Psychiatr Care*. 2019;55(3):401–8. https://doi.org/10.1111/ppc.12324

105 Diamond K, Mowszowski L, Cockayne N, et al. Randomized controlled trial of a healthy brain ageing cognitive training program: effects on memory, mood, and sleep. *J Alzheimers Dis*. 2015;44(4):1181–91. https://doi.org/10.3233/jad-142061

106 de Almondes KM, Leonardo MEM, Moreira AMS. Effects of a cognitive training program and sleep hygiene for executive functions and sleep quality in healthy elderly. *Dement Neuropsychol*. 2017;11(1):69–78. https://doi.org/10.1590/1980-57642016dn11-010011

107 Ladenbauer J, Külzow N, Passmann S, et al. Brain stimulation during an afternoon nap boosts slow oscillatory activity and memory consolidation in older adults. *Neuroimage*. 2016;142:311–23. https://doi.org/10.1016/j.neuroimage.2016.06.057

108 Westerberg CE, Florczak SM, Weintraub S, et al. Memory improvement via slow-oscillatory stimulation during sleep in older adults. *Neurobiol Aging*. 2015;36(9):2577–86. https://doi.org/10.1016/j.neurobiolaging.2015.05.014

109 Ladenbauer J, Ladenbauer J, Külzow N, et al. Promoting sleep oscillations and their functional coupling by transcranial stimulation enhances memory consolidation in mild cognitive impairment. *J Neurol Sci*. 2017;37(30):7111–24. https://doi.org/10.1523/JNEUROSCI.0260-17.2017

110 Jones AP, Choe J, Bryant NB, et al. Dose-dependent effects of closed-loop tACS delivered during slow-wave oscillations on memory consolidation. Original research. *Front Neurosci*. 2018;12. https://doi.org/10.3389/fnins.2018.00867

111 Massimini M, Ferrarelli F, Esser SK, et al. Triggering sleep slow waves by transcranial magnetic stimulation. *Proc Natl Acad Sci U S A*. 2007;104(20):8496–501. https://doi.org/10.1073/pnas.0702495104

112 Papalambros NA, Santostasi G, Malkani RG, et al. Acoustic enhancement of sleep slow oscillations and concomitant memory improvement in older adults. *Front Hum Neurosci*. 2017;109. https://doi.org/10.3389/fnhum.2017.00109

113 Papalambros NA, Weintraub S, Chen T, et al. Acoustic enhancement of sleep slow oscillations in mild cognitive impairment. *Ann Clin Transl Neurol*. 2019;6(7):1191–201. https://doi.org/10.1002/acn3.796

114 Proctor A, Bianchi MT. Clinical pharmacology in sleep medicine. *ISRN Pharmacol*. 2012;2012:914168. https://doi.org/10.5402/2012/914168

115 He Q, Chen X, Wu T, Li L, Fei X. Risk of dementia in long-term benzodiazepine users: evidence from a meta-analysis of observational studies. *J Clin Neurol*. 2019;15(1):9–19. https://doi.org/10.3988/jcn.2019.15.1.9

116 Kaestner EJ, Wixted JT, Mednick SC. Pharmacologically increasing sleep spindles enhances recognition for negative and high-arousal memories. *J Cogn Neurosci*. 2013;25(10):1597–610. https://doi.org/10.1162/jocn_a_00433

117 Tapiainen V, Taipale H, Tanskanen A, Tiihonen J, Hartikainen S, Tolppanen AM. The risk of Alzheimer's disease associated with benzodiazepines and related drugs: a nested case-control study. *Acta Psychiatr Scand*. 2018;138(2):91–100. https://doi.org/10.1111/acps.12909

118 Biétry FA, Pfeil AM, Reich O, Schwenkglenks M, Meier CR. Benzodiazepine use and risk of developing Alzheimer's disease: a case-control study based on swiss claims data. *CNS Drugs*. 2017;31(3):245–51. https://doi.org/10.1007/s40263-016-0404-x

119 Xue T, Wu X, Chen S, et al. The efficacy and safety of dual orexin receptor antagonists in primary insomnia: a systematic review and network meta-analysis. *Sleep Med Rev*. 2022;61:101573. https://doi.org/10.1016/j.smrv.2021.101573

120 Zhou F, Yan X-D, Wang C, et al. Suvorexant ameliorates cognitive impairments and pathology in APP/PS1 transgenic mice. *Neurobiol Aging.* 2020;91:66–75. https://doi.org/10.1016/j.neurobiolaging.2020.02.020

121 Muehlan C, Vaillant C, Zenklusen I, Kraehenbuehl S, Dingemanse J. Clinical pharmacology, efficacy, and safety of orexin receptor antagonists for the treatment of insomnia disorders. *Expert Opin Drug Metab Toxicol.* 2020;16(11):1063–78. https://doi.org/10.1080/17425255.2020.1817380

122 Ferracioli-Oda E, Qawasmi A, Bloch MH. Meta-analysis: melatonin for the treatment of primary sleep disorders. *PLoS ONE.* 2013;8(5):e63773. https://doi.org/10.1371/journal.pone.0063773

123 Xu J, Wang L-L, Dammer EB, et al. Melatonin for sleep disorders and cognition in dementia: a meta-analysis of randomized controlled trials. *Am J Alzheimer's Dis Other Dement.* 2015;30(5):439–47. https://doi.org/10.1177/1533317514568005

124 Furio AM, Brusco LI, Cardinali DP. Possible therapeutic value of melatonin in mild cognitive impairment: a retrospective study. *J Pineal Res.* 2007;43(4):404–9. https://doi.org/10.1111/j.1600-079X.2007.00491.x

125 Jean-Louis G, von Gizycki H, Zizi F. Melatonin effects on sleep, mood, and cognition in elderly with mild cognitive impairment. *J Pineal Res.* 1998;25(3):177–83. https://doi.org/10.1111/j.1600-079x.1998.tb00557.x

126 Cho JH, Bhutani S, Kim CH, Irwin MR. Anti-inflammatory effects of melatonin: a systematic review and meta-analysis of clinical trials. *Brain Behav Immun.* 2021;93:245–53. https://doi.org/10.1016/j.bbi.2021.01.034

127 Morvaridzadeh M, Sadeghi E, Agah S, et al. Effect of melatonin supplementation on oxidative stress parameters: a systematic review and meta-analysis. *Pharmacol Res.* 2020;161:105210. https://doi.org/10.1016/j.phrs.2020.105210

6

UNDERSTANDING THE COMPLEX LINK BETWEEN OBSTRUCTIVE SLEEP APNOEA AND CLINICAL DEPRESSION

Risk factors, mechanisms and effects of treatment

Melinda L. Jackson, Ivana Rosenzweig, Romola S. Bucks and Genevieve Rayner

Introduction

Obstructive sleep apnoea (OSA) is a common sleep disorder, estimated to affect almost one billion people worldwide (1). Recent studies estimate a prevalence of 23.4% in women and 49.7% in men over 40 years (2). It is a chronic disorder, characterised by repetitive collapse of the pharyngeal airway during sleep producing intermittent hypoxia and sleep disruption. The repetitive sleep disruption caused by untreated OSA has significant adverse consequences for daytime functioning, including profound daytime sleepiness (3) and impairments in cognitive function, mood and quality of life (4–6).

Depression is one of the most significant comorbidities seen in OSA (7, 8) and is of major concern given that comorbid depression is associated with poor adherence and compliance to OSA therapies (9), reduced quality of life (10), increased risk of suicide ideation (11) and poorer medical outcomes. The relationship between these conditions is complex. Some authors have argued that the psychiatric relevance of depressive symptoms in OSA patients has been overemphasised (12). Conversely, depression may be under-recognised in clinical practice. This chapter provides an overview of the prevalence, symptoms and risk factors for depression in OSA populations, as well as the frequency of antidepressant use among OSA patients. It then reviews the literature examining the impact of OSA treatment on depressive symptoms and clinical depression. Potential mechanisms and factors underpinning the link between OSA and depression are explored.

DOI: 10.4324/9781003296966-6

Prevalence of depression in OSA samples and issue of bidirectionality

Significantly higher rates of self-reported depressive symptoms (13–16) and clinician-diagnosed clinical depression (17, 18) have been identified in OSA patients relative to healthy controls. For instance, meta-analyses report between 17 and 45% of OSA patients have comorbid major depressive disorder (MDD) based on the Diagnostic and Statistical Manual of Mental Disorders (DSM-IV) criteria (10, 19), compared to 12-month prevalence rates of 6.6% in older adults from the broader community (20). A recent meta-analysis by Jackson and colleagues reported a pooled prevalence of 23% of clinical depression in OSA patients across seven studies (10). While not as common, OSA is also associated with elevated rates of psychopathologies other than depression, including anxiety, schizophrenia, posttraumatic stress disorder, substance abuse and bipolar disorder (for a review see (21)).

Stemming from the significant rates of depression in OSA patients, antidepressant use is common in people attending the sleep laboratory. Recent Australian studies report that 21–24.8% of patients were currently using antidepressants (10, 15). Similarly, Schwartz et al. found that 39% of 114 consecutive patients referred to the sleep laboratory were receiving antidepressant medication at the time of referral (22). Thus, not only is antidepressant use high in sleep clinic populations but also a significant proportion of OSA patients whose sleep disorder is untreated are using pharmacotherapy yet have residual depressive symptoms.

A point of uncertainty is the time course of comorbid depression and OSA; that is, what typically comes first, the depression or the OSA. There is evidence that OSA contributes to the causal pathway that leads to the expression of a depressive episode. Indeed, epidemiological studies using longitudinal data from people with and without OSA have identified OSA as an independent risk factor for depression. For example, in an observational community study of over 1,400 individuals, OSA was associated with a 1.8 fold increased risk of the onset of depression at four-year follow-up (23). A dose-response effect was observed for the odds of having a MDD diagnosis, with a 2.6-fold (95% CI: 1.7–3.9) increased risk of MDD in those with moderate to severe OSA (23). Similarly, Chen et al. (2013) compared incidence rates of depressive disorder in a sample of 2,818 OSA patients and 14,090 non-OSA controls during a one-year follow-up period (24). Within one year of diagnosis, the incidence of MDD was almost double among patients with OSA than in those without. In patients with OSA, there was also an independent increased risk (2.18 times) of consequent depressive disorder. These studies highlight the importance of treating sleep potentially to prevent the development, or reduce the severity, of later mental health issues.

Conversely, in some patients, the presence of depression may precipitate or exacerbate OSA through weight gain, decreased motivation for activity and exercise or sleep disruption. In others, both conditions may appear independently, though it is difficult to be certain that they arrive simultaneously as it is often unclear when OSA symptoms first manifest versus when they are first recognised. Development of scales better to capture information about when symptoms commenced and the length of time an individual has experienced OSA will provide a clearer understanding of the consequences of OSA on psychological and medical conditions. A promising new algorithm has been developed for estimating the onset of OSA from BMI and AHI (25), although further evidence is needed of its utility in practice. Moreover, more community- or population-based longitudinal studies will shed light on the temporal relationships between depression and OSA, and thus clarify whether one condition is a risk factor for the other.

Shared neuropathology, symptoms and risk factors of OSA and MDD

The relationship between OSA and depression is complex and not well-established, however, some explanations have been proposed to explain the high, but variable, comorbidity between these conditions. One possibility is that the association between OSA and depression is due to common risk factors, such as physical inactivity and alcohol use. Obesity is a condition common to both OSA and depression, which adds another layer to this complex relationship (26). Figure 6.1 depicts the potential bidirectional relationships between OSA and other comorbid conditions, including obesity and depression and the proposed underlying mechanisms (27).

Another possibility is that the association between OSA and depression might be spurious, instead due to the overlapping nature of signs and symptoms that are common to both conditions, such as lack of energy, disrupted sleep and poor concentration. In this case, the 'depressive symptoms' would be due to OSA and not to a depressive disorder (false positive cases of depression). Thus, there is debate as to whether depression in OSA is always a distinct clinical phenomenon, or could be a result of overlapping somatic and physical symptoms. Self-report scales of depression that are commonly used in research studies of OSA, such as the Beck Depression Inventory (BDI) and the Centre for Epidemiological Studies Depression Scale (CES-D), contain items relating to symptoms that are common to both OSA and depression (e.g., fatigue, somatic symptoms, loss of interest and poor concentration) (28). The use of these scales often makes it difficult to determine the independent presence of depression in OSA (7), and measures with greater symptom overlap produce higher estimates of depression prevalence in OSA (29). Indeed, prevalence estimates of depression in OSA can vary from as low as 8% to as high as 68% as the proportion of overlapping symptoms increases (29). However, there are depressive symptoms that do not appear to overlap markedly with OSA including negative affect, anhedonia and depressive cognitions (e.g., suicidal ideation, hopelessness), which can be used to differentiate depression from OSA symptoms. Some of the cardinal affective symptoms of depression, such as dysphoria (low mood) and anhedonia (a loss of interest or pleasure in things that normally bring joy), are not present in OSA, supporting the argument that major depression and OSA are separate conditions. Bucks and colleagues compared overlapping (e.g., insomnia, lethargy, impaired concentration, psychomotor retardation) and non-overlapping (e.g., agitation, loss of weight and appetite, negative affect, anhedonia and hypochondriasis) symptoms on the Hamilton Rating Scale for Depression (HAM-D) (29). A significantly greater proportion of patients endorsed overlapping (30%) compared to non-overlapping depressive symptoms (12%), consistent with the idea that the expression of OSA is a significant contributor to depressive symptomatology.

Since there are distinct features of depression that are independent of OSA, it is possible to diagnose depression in a patient with OSA with careful clinical assessment requiring research diagnostic evaluation that takes account of comorbid disorders diagnosing depression, a measurement that has rarely been conducted in research studies (discussed in a subsequent section below), and not feasible to conduct clinically in sleep laboratories. Within the reach of sleep laboratories is using depression scales with few overlapping symptoms such as the HADS or the Geriatric Depression Scale. Further understanding of the potential overlap between OSA and depression has been derived from OSA treatment studies.

Effect of treatment of OSA on depressive symptoms

Continuous positive airway pressure (CPAP) is the most efficacious (30) and cost-effective (31) treatment for OSA. CPAP prevents pharyngeal airway collapse during sleep, thereby improving sleep quality and oxygen saturation. While there is strong evidence demonstrating that CPAP therapy improves cognitive dysfunction in OSA patients (16, 32), there is a paucity of studies examining the effect of CPAP treatment on depressive symptoms (for a review see (7)). However, there is conflicting evidence regarding improvements in mood with CPAP use in the literature. Some report a reduction in mild depressive symptoms after up to six months of CPAP treatment (16, 22, 33–37), whereas others have observed no significant amelioration of depressive symptoms after up to one year of CPAP therapy (38, 39). A recent meta-analysis reported that CPAP did not improve symptoms of depression or anxiety (40). In contrast, another meta-analysis reported that CPAP improved depressive symptoms, but there was no treatment effect on anxiety (41).

One limitation that these studies do not address is the presence of overlapping and non-overlapping symptoms of depression and OSA inherent in many depression scales. In a unique study exploring the impact of OSA treatment on overlapping and non-overlapping symptoms as measured by a clinician-administered depression scale (Hamilton Rating Scale for Depression), Bucks et al. (2018) examined the impact of three months of CPAP on distinct symptoms of OSA. Whilst CPAP improved both types of symptoms, the overlapping symptoms showed the greatest improvement (29). This highlights the potential benefit of CPAP for alleviating the burden of OSA-related depressive symptoms, and further, that detection of symptoms that do not overlap with OSA may help to more accurately diagnose clinical depression in these patients.

Inadequate methodologies further limit the interpretation of many studies to date. In particular, many studies assess mood changes after just one to three weeks of CPAP treatment (42–46), when four to six weeks of treatment is typically required before differences in depression can be observed between the control and placebo groups due to (i) early placebo effects in both groups and (ii) the persistent nature of depressive symptoms. There are limited data on the efficacy of longer periods of CPAP therapy (>four weeks) for improving ratings on depression scales (13, 29, 45, 47–49). Similarly, there is a paucity of data on the effectiveness of mandibular splints for improving depressive symptoms (13, 50). Where they exist, these studies typically fail to find a treatment effect on depression ratings or find improvements in both CPAP and placebo arms of the trial, perhaps due to treatment improving depressive symptoms that are also symptoms of OSA.

A major limitation of previous research is that most of these studies were not specifically designed to examine the impact of CPAP on clinical depression. Many studies specifically exclude OSA patients meeting criteria for MDD and/or those with self-report depression scores outside the 'normal' range at baseline. Typically, self-report measures of depression were added ad hoc to the designs, thereby reducing the ability clearly to distinguish features of depression from OSA symptoms. The impact of CPAP treatment on mood may also depend on the severity of depression experienced. Many of the scales used to assess depression have not been validated in sleep-disordered populations, which is problematic given the overlap in symptomatology. Further evidence on the impact of OSA treatment on depression has been observed through studies in patients with clinical depression.

Effect of OSA treatment on major depressive disorder

While there is strong evidence that treatment of OSA improves depressive symptoms, there are limited studies examining whether treatment of OSA in OSA patients with comorbid major depression improves mood/psychiatric outcomes. The assessment of clinical depression via structured clinical interviews has rarely been conducted in OSA studies. The use of the gold-standard Structured Clinical Interview for DSM-5 (SCID), which detects the presence of depressive disorders according to the criteria of the Diagnostic and Statistical Manual for Mental Disorders, allows for a diagnosis of depression based on the presence of cardinal symptoms of depression (anhedonia, dysphoria) that do not overlap with OSA, mitigating against inflated diagnoses.

A number of uncontrolled studies have reported changes in depression diagnosis in clinically depressed OSA patients after CPAP (51–53) or surgical treatment to remove excess tissue in the throat to widen the airway (uvulopalatopharyngoplasty; UPPP; (54)). Three studies used the SCID both prior to and after two (51–53) or six (54) months of treatment. In all these studies, around a third of the sample had clinical depression at baseline (29.7%–34%). At follow-up, the depression of half of these participants had remitted (i.e., they no longer met criteria for any depressive disorder). Six months after UPPP surgery, the percentage of participants who met criteria for major depression dropped from 34% to 10%, and no new cases of depression were reported (54). Another study used the HAM-D to track the trajectory of depressive symptoms in a sample of 17 OSA patients with clinically diagnosed MDD who were using antidepressants but had residual depression symptoms (53). A significant reduction in HAM-D scores was observed after 2 months of CPAP therapy, however, antidepressant use at follow-up was not reported.

The only controlled trial, to our knowledge, was conducted in a clinical sample of 121 OSA patients who were assessed before and after CPAP using the SCID (55). At four-month follow-up, those who were randomised to receive CPAP therapy had a significant reduction in clinical depression (from 20.5% to 7%) compared to a waitlist control group (from 24% to 35%). Importantly, five patients in the CPAP group ceased using antidepressant medication at follow-up, and no longer met criteria. This is supported by other studies that have tracked rates of antidepressant use before and after CPAP treatment (56, 57). For example, in a sub-analysis of nine OSA patients taking antidepressant medication at baseline, Schwartz and Karatinos (2007) observed improvements in depressive symptoms on the BDI-FS after six weeks of CPAP therapy (from a score of 4.2–1.4) that was sustained after one year of CPAP treatment (57). Three of nine patients ceased using their antidepressants after commencing CPAP treatment. In a large observational study, 300 OSA patients were followed up after at least one year of CPAP use (56). Around 42% endorsed persistent depressive symptoms after >1 year of CPAP. Of the 89 participants who were using antidepressants at baseline, 46% had discontinued their medication by follow-up. Of these, antidepressant therapy was discontinued during follow-up in 14 (35%) with persistent depressive symptoms and 27 (55%) without persistent depressive symptoms while using CPAP. Thus, depression did not resolve in all patients after CPAP treatment and ceasing antidepressant therapy following CPAP was not always indicative of remission of depressive symptoms. Seventeen participants had been prescribed antidepressants during the follow-up period, suggesting that ~6% of the cohort met criteria for treatment during the CPAP intervention. These findings indicate that, while CPAP may be linked to a reduction in antidepressant use following the commencement of OSA treatment, this may not necessarily indicate remission of the mood disorder.

Clearly, there is a paucity of trials examining the impact of CPAP on mood in OSA patients with clinically significant depression. One of the main limitations of these studies is that they are open-label trials, examining depression before and after CPAP without an appropriate control group. A placebo effect is clearly observed in depression scores in past OSA intervention studies (13, 44), and thus may be impacting on the study outcomes. CPAP usage data are not reported or measured in all trials making it difficult to determine whether there is an association between CPAP use and improvements in mood, or indeed if CPAP adherence is an issue for patients with depression.

As has been reported in numerous studies of untreated OSA, persistent depression after CPAP is linked to excessive daytime sleepiness (EDS). Even in patients who are compliant with CPAP treatment, residual EDS is strongly positively associated with depression after CPAP use (56), with reduced EDS associated with improvements in depression after CPAP (53). Sleep fragmentation is considered a primary cause of EDS (58). While depression can play a role in non-adherence to treatment, a common symptom of depression is excessive fatigue which may be an underlying cause of persistent sleepiness following adequate treatment of OSA, requiring further investigation.

Potential factors that mediate the link between OSA and MDD

There are a number of neurophysiological and cognitive factors that may contribute to the excessive rates of depression seen in OSA, which will be explored further in this section, and depicted in Figure 6.1. The physiological consequences of repetitive collapse of the pharyngeal airway during sleep include sleep fragmentation and intermittent hypoxia/reoxygenation. Sleep fragmentation is a direct consequence of the repetitive arousals following apnoeas and hypopneas. Longitudinal data show that individuals experiencing sleep difficulties and insomnia have a two- to fourfold increased risk of developing new-onset depression years later, indicating that sleep disturbance in otherwise healthy individuals plays an etiological role in the development of depression (59, 60). Additionally, functional MRI studies have shown that sleep deprivation increases amygdala reactivity (a brain structure of the limbic system involved in emotional processing), thereby amplifying responses to negative emotional stimuli (61). Thus, chronic sleep disruption in OSA may be a precipitating factor that 'triggers' depressed mood in these patients.

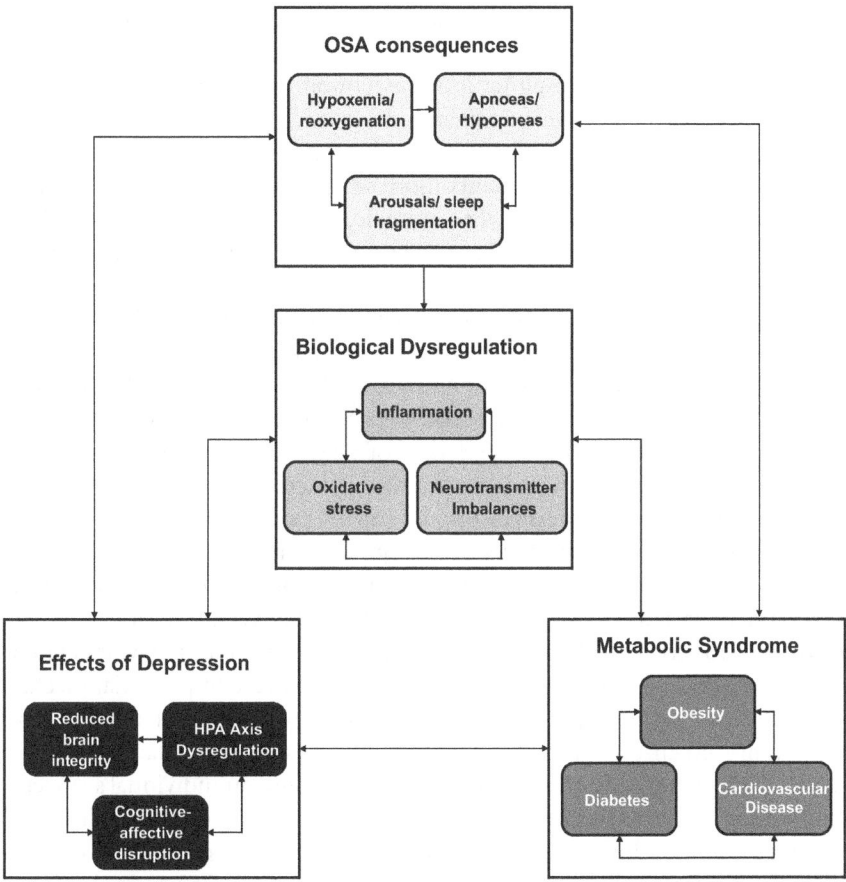

FIGURE 6.1 Proposed bidirectional mechanisms of OSA pathophysiology, biological dysregulation and comorbidities linking obstructive sleep apnoea
and depression. Each cluster is an independent entry point to the
cycle. If left untreated, the presence of a risk factor increases the likelihood of the synergistic development of more symptoms from each
cluster, increasing the risk of depression and/or OSA.

It has been argued that in some patients with OSA, fragmented REM sleep
may precipitate a vicious cycle of impaired regulation and rebound of REM,
which, along with concomitant changes in neurotransmitter systems caused by
hypoxemia, could further reduce monoamine activity (62). This chemical milieu
could potentially lead to increased negative rumination and ensuing depression,
particularly in patients with are genetically predisposed to mood disturbance.
Fragmented REM sleep has also been linked with an increase in negative affect,
a reduction in positive affect, and greater difficulty regulating one's emotional
responses (63). Together, this evidence raises the question as to whether the

high rates of depression reported in the OSA population, who experience significant sleep disturbance, are related to poor emotional regulation and heightened emotional reactivity resulting from disruption of sleep microarchitecture. Indeed, OSA is associated with significant changes in cognitive function (6, 64, 65), including social and emotional impairments (64, 66). Thus, these neurocognitive changes may make patients more susceptible to depression, or may manifest as depressive symptoms. For example, the severity of reported MDD symptoms has been linked to an over-reliance on maladaptive emotional regulation strategies (such as rumination, suppression, avoidance) and limited use of adaptive emotion regulation strategies (such as cognitive reappraisal, mindfulness, reframing) (67). Depressed individuals also have greater difficulty implementing adaptive emotional regulation strategies than non-depressed individuals (68). For instance, the response style theory proposes that depressed individuals are more likely to respond to experiences of negative affect with rumination, which in turn exacerbates their depressed mood, increases their negative thinking and predicts future depressive episodes (69).

One concept of late-onset depression posits it to be precipitated by an increasing burden of ischemic lesions, undermining the integrity of white matter tracts and disconnecting communication between key nodes in the limbic-cortico-striato-pallido-thalamic network (70). Changes in white matter integrity also occur in OSA (71–73) and are therefore one potential mechanism for the onset of depression in untreated OSA patients. It is likely that other comorbid chronic diseases, including obesity, cardiovascular disease and diabetes, may be driving this effect, thereby increasing an individual's susceptibility to depression. Indeed, cardiovascular risk factors are the key driver of the white matter lesion burden theorised to cause the depression (74). Sleep is an important process during which myelination occurs (75), and oligodendrocytes, the myelin-producing cells in the CNS, are sensitive to sleep fragmentation (76). The presence of comorbid conditions may also explain why depression does not resolve in some patients after treatment of OSA. While OSA in and of itself impacts on cognitive functioning (64) and markers of brain integrity, the contributions of comorbid conditions alongside OSA need to be further examined in future studies.

Neuroimaging studies of depressed individuals have reported reductions in grey matter in regions that overlap with those observed in OSA individuals, such as the hippocampus, amygdala and regions of the frontal cortex (77). In addition, reduced grey matter volume in regions associated with affective processing, including the cerebellum is observed in OSA (78), as well as brain metabolite changes that are found in frontal and hippocampal regions compared to healthy controls may increase patients' susceptibility to depression (16, 79). In support of this theory, grey matter volume differences have been observed between OSA patients with and without depression (80). Frontal and hippocampal damage is associated with memory impairment, and feelings of worthlessness, guilt, sadness and suicidality, which are also cognitive aspects of depression (77). Thus, early treatment of OSA may reduce neural damage to these brain regions, and potentially the progression of depression.

Inflammatory processes are also likely to be driving the development and pathogenesis of depression in OSA patients. It is well-recognised that major depression is associated with an immune-inflammatory response (81), via an increase in the brain's sensitivity to stress (82). In particular, altered functioning of neural networks in individuals with depression has been associated with increased circulating cytokines, leading to dysregulation of synaptic plasticity in the neural pathways linked with depression (83). Indeed, increased serum inflammatory cytokines have been observed in OSA patients (84). While an adaptive mechanism of intermittent hypoxia has been argued to include ischemic preconditioning (85), a maladaptive feature of this symptom is neuroinflammation. Interestingly, intermittent hypoxia exposure in animal models of OSA has been previously associated with an early antidepressogenic effect, as well as to an inflammatory signature in the distinct fronto-brainstem subcircuitry that includes the serotonergic dorsal raphe nucleus (71). The activation of a similar brain circuitry has been also reported to increase effortful behavioural responses to challenging situations (86). It has been theorised that whilst this behavioural response may, initially, have an adaptive and beneficial role, it is simultaneously metabolically very demanding, and, in situations where there is no immediate resolution to the challenge, this behaviour may lead to an increased anxiety and depression (71, 87, 88). A mixed anxiety and depression endophenotype is known to occur in patients with OSA (89) and has been linked with higher suicide risks in depressed patients (71, 90). Of note, in postmortem brains of depressed patients who committed suicide, an increased inflammatory signal has been found in frontal cortical regions (91). Thus, whilst speculative, it is conceivable that in a subgroup of OSA patients, early induction of the neuroinflammatory process in frontal regions, although initially adaptive, may subsequently lead to heightened risk of developing a specific, complex endophenotype of depression with high anxiety component.

In keeping with this and other preclinical work, alterations in both central and peripheral serotonin have long been implicated in the pathogenesis of major depression (92). Serotonergic function has also been shown to influence upper-airway dilator muscle activity during sleep (93). Changes to serotonin levels lead to alterations in sleep architecture in depression (94), thus presenting another potential mechanism linking OSA and depression.

Summary

While previous literature suggests that depressive symptoms may improve with treatment of OSA, these studies have either not specifically included OSA patients with comorbid MDD, have used limited treatment periods or have not included a control group. There is preliminary evidence that treatment of OSA may alleviate psychiatric burden, however, controlled clinical trials of patients with OSA and clinical depression are needed to support this hypothesis. The few studies that have used a standardised clinical interview to diagnose current MDD have shown promising results, with around half of the participants not meeting the criteria for MDD at follow-up.

As expected, there are some patients who maintain their depression status after their OSA has been adequately treated. Future studies are needed to assess whether adjunct therapies for depression (e.g., cognitive behavioural therapy, antidepressants) may further benefit OSA patients with comorbid MDD who commence CPAP. In the clinical setting, these patients need additional follow-up and referral to mental health specialists for further assessment and treatment (95). The gap in our knowledge with regards to the efficacy of CPAP treatment for improving depression is clinically very important, and solid evidence is required to support clinicians' decisions to treat OSA patients who also have MDD, given that depression can impede OSA treatment adherence and cause significant burden to the patient.

References

1 Benjafield AV, Ayas NT, Eastwood PR, Heinzer R, Ip MSM, Morrell MJ, et al. Estimation of the global prevalence and burden of obstructive sleep apnoea: a literature-based analysis. Lancet Respir Med. 2019;7(8):687–98.

2 Heinzer R, Vat S, Marques-Vidal P, Marti-Soler H, Andries D, Tobback N, et al. Prevalence of sleep-disordered breathing in the general population: the HypnoLaus study. Lancet Respir Med. 2015;3(4):310–18.

3 Guilleminault C, Tilkian A, Dement WC. The sleep apnea syndromes. Ann Rev Med. 1976;27(1):465–84.

4 Ferini-Strambi L, Baietto C, Di Gioia MR, Castaldi P, Castronovo C, Zucconi M, et al. Cognitive dysfunction in patients with obstructive sleep apnea: partial reversibility after continuous positive airway pressure. Brain Res Bull. 2003;61(1):87–92.

5 Twigg GL, Papaioannou I, Jackson ML, Ghiassi R, Shaikh Z, Jaye J, et al. Obstructive sleep apnea syndrome is associated with deficits in verbal but not visual memory. AJRCCM. 2010;182(1):98–103.

6 Jackson ML, Howard ME, Barnes M. Cognition and daytime functioning in sleep-related breathing disorders. In: Van Dongen H, Kerkhof G, editors. Human Sleep and Cognition Progress in Brain Research. Oxford, UK: Elsevier; 2011. p. 53–68.

7 Harris M, Glozier N, Ratnavadivel R, Grunstein RR. Obstructive sleep apnea and depression. Sleep Med Rev. 2009;13(6):437–44.

8 Saunamäki T, Jehkonen M. Depression and anxiety in obstructive sleep apnea syndrome: a review. Acta Neurol Scand. 2007;116(5):277–88.

9 Law M, Naughton M, Ho S, Roebuck T, Dabscheck E. Depression may reduce adherence during CPAP titration trial. J Clin Sleep Med. 2014;10(2):163–9.

10 Jackson M, Tolson J, Bartlett D, Berlowitz DJ, Varma P, Barnes M. Clinical depression in untreated Obstructive Sleep Apnea: examining predictors and a meta-analysis of prevalence rates. Sleep Med. 2019;62:22–8.

11 Bishop TM, Ashrafioun L, Pigeon WR. The association between sleep apnea and suicidal thought and behavior: an analysis of national survey data. J Clin Psychiatry. 2017;79(1).

12 Lee S. Depression in sleep apnea: a different view. J Clin Psychiatry. 1990;51:309–10.

13 Barnes M, McEvoy RD, Banks S, Tarquinio N, Murray CG, Vowles N, et al. Efficacy of positive airway pressure and oral appliance in mild to moderate obstructive sleep apnea. Am J Respir Crit Care Med. 2004;170(6):656–64.

14 Jackson ML, Stough C, Howard ME, Spong J, Downey LA, Thompson B. The contribution of fatigue and sleepiness to depression in patients attending the sleep laboratory for evaluation of obstructive sleep apnea. Sleep Breath. 2011;15(3):439–45.

15 Law M, Naughton MT, Dhar A, Barton D, Dabscheck E. Validation of two depression screening instruments in a sleep disorders clinic. JClinSleepMed. 2014;10(6):683–8.

16 O'Donoghue FJ, Wellard RM, Rochford PD, Dawson A, Barnes M, Ruehland WR, et al. Magnetic resonance spectroscopy and neurocognitive dysfunction in obstructive sleep apnea before and after CPAP treatment. Sleep. 2012;35(1):41–8.

17 Ohayon MM. The effects of breathing-related sleep disorders on mood disturbances in the general population. J Clin Psychiatry. 2003;64(10):1195–200.

18 Sharafkhaneh A, Giray N, Richardson P, Young T, Hirshkowitz M. Association of psychiatric disorders and sleep apnea in a large cohort. Sleep. 2005;28(11):1405–11.

19 Edwards C, Almeida OP, Ford AH. Obstructive sleep apnea and depression: a systematic review and meta-analysis. Maturitas. 2020;142:45–54.

20 Mojtabai R, Olfson M. Major depression in community-dwelling middle-aged and older adults: prevalence and 2- and 4-year follow-up symptoms. Psychol Med. 2004;34(4):623–34.

21 Naqvi HA, Wang D, Glozier N, Grunstein RR. Sleep-disordered breathing and psychiatric disorders. Curr Psychiatr Rep. 2014;16(12):1–11.

22 Schwartz DJ, Kohler WC, Karatinos G. Symptoms of depression in individuals with obstructive sleep apnea may be amenable to treatment with continuous positive airway pressure. Chest. 2005;128(3):1304–9.

23 Peppard PE, Szklo-Coxe M, Hla KM, Young T. Longitudinal association of sleep-related breathing disorder and depression. Arch Intern Med. 2006;166(16):1709–15.

24 Chen Y-H, Keller JK, Kang J-H, Hsieh H-J, Lin H-C. Obstructive sleep apnea and the subsequent risk of depressive disorder: a population-based follow-up study. J Clin Sleep Med. 2013;9(5):417–23.

25 Olaithe M, Hagen EW, Barnet JH, Eastwood PR, Bucks RS. OSA-onset: an algorithm for predicting the age of OSA onset. Sleep Med. 2023;108:100–4.

26 Penney D, Barbera J. CPAP adherence in patients with obstructive sleep apnea and depression. In: Shapiro CM, Gupta M, Zalai D, editors. CPAP Adherence: Factors and Perspectives. Cham: Springer International Publishing; 2022. p. 203–11.

27 Gupta MA, Simpson FC. Obstructive sleep apnea and psychiatric disorders: a systematic review. J Clin Sleep Med. 2015;11(2):165–75.

28 Nanthakumar S, Bucks RS, Skinner TC. Are we overestimating the prevalence of depression in chronic illness using questionnaires? Meta-analytic evidence in obstructive sleep apnoea. Health Psychol. 2016;35(5):423.

29 Bucks RS, Nanthakumar S, Starkstein SS, Hillman DR, James A, McArdle N, et al. Discerning depressive symptoms in patients with obstructive sleep apnea: the effect of continuous positive airway pressure therapy on Hamilton Depression Rating Scale symptoms. Sleep. 2018;41(12).

30 McDaid C, Griffin S, Weatherly H, Durée K, van der Burgt M, van Hout S, et al. Continuous positive airway pressure devices for the treatment of obstructive sleep apnoea – hypopnoea syndrome: a systematic review and economic analysis. Health Technol Ass. 2009;13(4):143–274.

31 Deloitte Access Economics. Reawakening Australia: The Economic Cost of Sleep Disorders in Australia, 2010. Barton, Australia: Sleep Health Foundation; 2011.

32 Canessa N, Castronovo V, Cappa SF, Aloia MS, Marelli S, Falini A, et al. Obstructive sleep apnea: brain structural changes and neurocognitive function before and after treatment. Am J Respir Crit Care Med. 2011;183(10):1419–26.

33 Castronovo V, Canessa N, Strambi LF, Aloia MS, Consonni M, Marelli S, et al. Brain activation changes before and after PAP treatment in Obstructive Sleep Apnea. Sleep. 2009;32(9):1161–72.

34 Yamamoto H, Akashiba T, Kosaka N, Ito D, Horie T. Long-term effects nasal continuous positive airway pressure on daytime sleepiness, mood and traffic accidents in patients with obstructive sleep apnoea. Resp Med. 2000;94(1):87–90.

35 Means MK, Lichstein KL, Edinger JD, Taylor DJ, Durrence HH, Husain AM, et al. Changes in depressive symptoms after continuous positive airway pressure treatment for obstructive sleep apnea. Sleep Breath. 2003;7(1):31–42.

36 Sanchez AI, Buela-Casal G, Bermudez MP, Casas-Maldonado F. The effects of continuous positive air pressure treatment on anxiety and depression levels in apnea patients. Psychiatry Clin Neurosci. 2001;55(6):641–6.

37 Kingshott RN, Vennelle M, Hoy CJ, Engleman HM, Deary IJ, Douglas NJ. Predictors of improvements in daytime function outcomes with CPAP therapy. Am J Respir Crit Care Med. 2000;161(3):866–71.

38 Borak J, Cieslicki J, Koziej M, Matuszewski A, Zielinski J. Effects of CPAP treatment on psychological status in patients with severe obstructive sleep apnoea. J Sleep Res. 1996;5(2):123–7.

39 Munoz A, Mayoralas L, Barbe F, Pericas J, Agusti A. Long-term effects of CPAP on daytime functioning in patients with sleep apnoea syndrome. Euro Resp J. 2000;15(4):676–81.

40 Patil SP, Ayappa IA, Caples SM, Kimoff RJ, Patel SR, Harrod CG. Treatment of adult obstructive sleep apnea with positive airway pressure: an American Academy of Sleep Medicine systematic review, meta-analysis, and GRADE assessment. J Clin Sleep Med. 2019;15(2):301–34.

41 Zheng D, Xu Y, You S, Hackett ML, Woodman RJ, Li Q, et al. Effects of continuous positive airway pressure on depression and anxiety symptoms in patients with obstructive sleep apnoea: results from the sleep apnoea cardiovascular Endpoint randomised trial and meta-analysis. EClinicalMedicine. 2019;11:89–96.

42 Yu BH, Ancoli-Israel S, Dimsdale JE. Effect of CPAP treatment on mood states in patients with sleep apnea. J Psychiatric Res. 1999;33(5):427–32.

43 Bardwell WA, Ancoli-Israel S, Dimsdale JE. Comparison of the effects of depressive symptoms and apnea severity on fatigue in patients with obstructive sleep apnea: a replication study. J Affect Dis. 2007;97(1–3):181–6.

44 Haensel A, Norman D, Natarajan L, Bardwell WA, Ancoli-Israel S, Dimsdale JE. Effect of a 2 week CPAP treatment on mood states in patients with obstructive sleep apnea: a double-blind trial. Sleep Breath. 2007;11(4):239–44.

45 Henke KG, Grady JJ, Kuna ST. Effect of nasal continuous positive airway pressure on neuropsychological function in Sleep Apnea – Hypopnea Syndrome: a randomized, placebo-controlled trial. Am J Respir Crit Care Med. 2001;163(4):911–17.

46 Lee IS, Bardwell W, Ancoli-Israel S, Loredo JS, Dimsdale JE. Effect of three weeks of continuous positive airway pressure treatment on mood in patients with obstructive sleep apnoea: a randomized placebo-controlled study. Sleep Med. 2012;13(2):161–6.

47 Engleman H, Kingshott RN, Wraith PK, Mackay TW, Deary IJ, Douglas NJ. Randomized placebo-controlled crossover trial of continuous positive airway pressure for mild sleep apnea/hypopnea syndrome. AJRCCM. 1999;159(2):461–7.

48 Engleman H, Martin SE, Douglas NJ, Deary IJ. Effect of continuous positive airway pressure treatment on daytime function in sleep apnoea/hypopnoea syndrome. Lancet. 1994;343(8897):572–5.

49 Barnes M, Houston D, Worsnop CJ, Neill AM, Mykytyn IJ, Kay A, et al. A randomized controlled trial of continuous positive airway pressure in mild obstructive sleep apnea. AJRCCM. 2002;165(6):773–80.

50 Naismith S, Winter V, Hickie I, Cistulli P. Effect of oral appliance therapy on neurobehavioral functioning in obstructive sleep apnea: a randomized controlled trial. J Clin Sleep Med. 2005;1(4):374–80.

51 Eldahdouh SS, El-Habashy MM, Elbahy MS. Effect of CPAP on depressive symptoms in OSA. Egypt J Chest Dis Tuberc. 2014;63(2):389–93.

52 El-Sherbini AM, Bediwy AS, El-Mitwalli A. Association between obstructive sleep apnea (OSA) and depression and the effect of continuous positive airway pressure (CPAP) treatment. Neuropsychiatr Dis Treat. 2011;7:715–21.

53 Habukawa M, Uchimura N, Kakuma T, Yamamoto K, Ogi K, Hiejima H, et al. Effect of CPAP treatment on residual depressive symptoms in patients with major depression and coexisting sleep apnea: contribution of daytime sleepiness to residual depressive symptoms. Sleep Med. 2010;11(6):552–7.

54 Dahlof P, Ejnell H, Hallstrom T, Hedner J. Surgical treatment of the sleep apnea syndrome reduces associated major depression. Int J Behav Med. 2000;7(1):73–88.

55 Jackson ML, Tolson J, Schembri R, Bartlett D, Rayner G, Lee VV, et al. Does continuous positive airways pressure treatment improve clinical depression in obstructive sleep apnea? A randomized wait-list controlled study. Depress Anxiety. 2020;38(5):498–507.

56 Gagnadoux F, Le Vaillant M, Goupil F, Pigeanne T, Chollet S, Masson P, et al. Depressive symptoms before and after long-term CPAP therapy in patients with sleep apnea. Chest. 2014;145(5):1025–31.

57 Schwartz DJ, Karatinos G. For individuals with obstructive sleep apnea, institution of CPAP therapy is associated with an amelioration of symptoms of depression which is sustained long term. J Clin Sleep Med. 2007;3(6):631.

58 Sforza E, de Saint Hilaire Z, Pelissolo A, Rochat T, Ibanez V. Personality, anxiety and mood traits in patients with sleep-related breathing disorders: effect of reduced daytime alertness. Sleep Med. 2002;3(2):139–45.

59 Baglioni C, Battagliese G, Feige B, Spiegelhalder K, Nissen C, Voderholzer U, et al. Insomnia as a predictor of depression: a meta-analytic evaluation of longitudinal epidemiological studies. J Affective Dis. 2011;135(1–3):10–19.

60 Jackson ML, Sztendur EM, Diamond NT, Byles JE, Bruck D. Sleep difficulties and the development of depression and anxiety: a longitudinal study of young Australian women. Arch Wom Ment Health. 2014;17(3):189–98.

61 Yoo SS, Gujar N, Hu P, Jolesz FA, Walker MP. The human emotional brain without sleep – a prefrontal amygdala disconnect. Curr Biol. 2007;17(20):R877–8.

62 Rosenzweig I, Weaver TE, Morrell MJ. Obstructive sleep apnea and the central nervous system: neural adaptive processes, cognition, and performance. In: Kryger M, Roth T, Dement WC, editors. Principles and Practice of Sleep Medicine (Sixth Edition). Philadelphia, PA: Elsevier; 2017. p. 1154–66.e5.

63 Palmer CA, Alfano CA. Sleep and emotion regulation: an organizing, integrative review. Sleep Med Rev. 2017;31:6–16.

64 Gnoni V, Mesquita M, O'Regan D, Delogu A, Chakalov I, Antal A, et al. Distinct cognitive changes in male patients with obstructive sleep apnoea without co-morbidities. Front Sleep. 2023;2.

65 Bubu OM, Andrade AG, Umasabor-Bubu OQ, Hogan MM, Turner AD, de Leon MJ, et al. Obstructive sleep apnea, cognition and Alzheimer's disease: a systematic review integrating three decades of multidisciplinary research. Sleep Med Rev. 2020;50:101250.

66 Lee VV, Trinder J, Jackson ML. Autobiographical memory impairment in obstructive sleep apnea patients with and without depressive symptoms. J Sleep Res. 2016;25(5):605–11.

67 Fairholme CP, Manber R. Sleep, emotions, and emotion regulation: an overview. Sleep Affect. 2015:45–61.

68 Ehring T, Tuschen-Caffier B, Schnülle J, Fischer S, Gross JJ. Emotion regulation and vulnerability to depression: spontaneous versus instructed use of emotion suppression and reappraisal. Emotion. 2010;10(4):563–72.

69 Nolen-Hoeksema S, Wisco BE, Lyubomirsky S. Rethinking rumination. Perspect Psychol Sci. 2008;3(5):400–24.

70 Alexopoulos GS, Young RC, Meyers BS, Abrams RC. Late-onset depression. Psychiatr Clin North Am. 1988;11(1):101–15.

71 Polsek D, Cash D, Veronese M, Ilic K, Wood TC, Milosevic M, et al. The innate immune toll-like-receptor-2 modulates the depressogenic and anorexiolytic neuroinflammatory response in obstructive sleep apnoea. Sci Rep. 2020;10(1):11475.

72 Castronovo V, Scifo P, Castellano A, Aloia MS, Iadanza A, Marelli S, et al. White matter integrity in obstructive sleep apnea before and after treatment. Sleep. 2014;37(9):1465–75.

73 Lee M-H, Lee SK, Kim S, Kim REY, Nam HR, Siddiquee AT, et al. Association of obstructive sleep apnea with white matter integrity and cognitive performance over a 4-year period in middle to late adulthood. JAMA Netw Open. 2022;5(7):e2222999-e.

74 Alexopoulos GS, Meyers BS, Young RC, Campbell S, Silbersweig D, Charlson M. 'Vascular depression' hypothesis. Arch Gen Psychiatry. 1997;54(10):915–22.

75 de Vivo L, Bellesi M. The role of sleep and wakefulness in myelin plasticity. Glia. 2019;67(11):2142–52.

76 Bellesi M, Pfister-Genskow M, Maret S, Keles S, Tononi G, Cirelli C. Effects of sleep and wake on oligodendrocytes and their precursors. J Neurosci. 2013;33(36):14288–300.

77 Drevets WC. Neuroimaging and neuropathological studies of depression: implications for the cognitive-emotional features of mood disorders. Curr Opin Neurobiol. 2001;11(2):240–9.

78 Morrell MJ, Jackson ML, Twigg GL, Ghiassi R, McRobbie DW, Quest RA, et al. Changes in brain morphology in patients with obstructive sleep apnoea. Thorax. 2010;65(10):908–14.

79 Bartlett DJ, Rae C, Thompson CH, Byth K, Joffe DA, Enright T, et al. Hippocampal area metabolites relate to severity and cognitive function in obstructive sleep apnea. Sleep Med. 2004;5(6):593–6.

80 Cross RL, Kumar R, Macey PM, Doering LV, Alger JR, Yan-Go FL, et al. Neural alterations and depressive symptoms in obstructive sleep apnea patients. Sleep. 2008;31(8):1103.

81 Howren MB, Lamkin DM, Suls J. Associations of depression with C-reactive protein, IL-1, and IL-6: a meta-analysis. Psychosom Med. 2009;71(2):171–86.

82 Skaper S, Facci L, Giusti P. Neuroinflammation, microglia and mast cells in the pathophysiology of neurocognitive disorders: a review. CNS Neurolo Disord Drug Targets. 2014;13(10):1654–66.

83 Piser TM. Linking the cytokine and neurocircuitry hypotheses of depression: a translational framework for discovery and development of novel anti-depressants. Brain Behav Immun. 2010;24(4):515–24.

84 Ciftci TU, Kokturk O, Bukan N, Bilgihan A. The relationship between serum cytokine levels with obesity and obstructive sleep apnea syndrome. Cytokine. 2004;28(2):87–91.

85 Rosenzweig I, Kempton MJ, Crum WR, Glasser M, Milosevic M, Beniczky S, et al. Hippocampal hypertrophy and sleep apnea: a role for the ischemic preconditioning? PLoS ONE. 2013;8(12):e83173.

86 Warden MR, Selimbeyoglu A, Mirzabekov JJ, Lo M, Thompson KR, Kim S-Y, et al. A prefrontal cortex – brainstem neuronal projection that controls response to behavioural challenge. Nature. 2012;492(7429):428–32.

87 Mu Y, Bennett DV, Rubinov M, Narayan S, Yang CT, Tanimoto M, et al. Glia accumulate evidence that actions are futile and suppress unsuccessful behavior. Cell. 2019;178(1):27–43e19.

88 Gonzalez-Torres ML, Dos Santos CV. Uncontrollable chronic stress affects eating behavior in rats. Stress. 2019;22(4):501–8.

89 Rezaeitalab F, Moharrari F, Saberi S, Asadpour H, Rezaeetalab F. The correlation of anxiety and depression with obstructive sleep apnea syndrome. J Res Med Sci. 2014;19(3):205–10.

90 Sareen J, Cox BJ, Afifi TO, de Graaf R, Asmundson GJ, ten Have M, et al. Anxiety disorders and risk for suicidal ideation and suicide attempts: a population-based longitudinal study of adults. Arch Gen Psychiatry. 2005;62(11):1249–57.

91 Pandey GN, Rizavi HS, Bhaumik R, Ren X. Innate immunity in the postmortem brain of depressed and suicide subjects: role of Toll-like receptors. Brain Behav Immun. 2019;75:101–11.

92 Maes M, Meltzer H. The serotonin hypothesis of major depression. In: Bloom F, Kupher D, editors. Psychopharmacology: The Fourth Generation of Progress. New York: Raven Press; 1995. p. 933–4.

93 Fenik P, Veasey SC. Pharmacological characterization of serotonergic receptor activity in the hypoglossal nucleus. Am J Respir Crit Care Med. 2003;167(4):563–9.

94 Schroder C, O'Hara R. Depression and obstructive sleep apnea (OSA). Ann Gen Psychiatry. 2005;4(13):1–8.

95 Beck A, Steer R, Brown G. Beck Depression Inventory-II Manual. San Antonio: The Psychological Society; 1996.

7

THE IMPACT OF INSUFFICIENT SLEEP ON COGNITIVE AND EMOTIONAL HEALTH IN ADOLESCENCE

Current advances and research needs

Gina M. Mason and Jared M. Saletin

Acknowledgements: This manuscript was supported in part by research grants from NIH (R01HD103655 to JMS and P20GM139743). We thank Victoria Dionisos and Taylor Christiansen for their helpful feedback during the preparation of this work.

Introduction

Adolescence is a tumultuous time of biological and psychosocial development (1). In the midst of growth and new experiences, teens are exposed to a pernicious near-nightly loss of sleep. Here, we review the science of insufficient sleep in adolescents. We describe a) how insufficient sleep arises during puberty; b) how sleep loss may impact waking behaviour with an emphasis on learning, attention, and emotion; and c) how societal factors might moderate these effects. Though work remains to be done, one thing is clear: while sleep loss is endemic amongst teenagers, neither it nor its consequences need be predetermined.

The perfect storm: one now capable of being predicted

Across the globe, adolescents lose upwards of 10–15 hours of sleep a week (2). Any discussion of this sleep loss begins with the societal pressures we place on developing teens: namely, school, academic demands and social and athletic obligations. If a teen describes a preference for staying up later at night and sleeping in later each morning they risk being stereotyped as 'lazy'. These prejudicial explanations of adolescent sleep have inhibited a compassionate understanding rooted in the fundamental biology of sleep and circadian rhythms. Over the past three decades, however, these interacting processes are now

DOI: 10.4324/9781003296966-7

understood to create a 'perfect storm' through which insufficient sleep can arise (3, 4; Figure 7.1). Two fundamental shifts occur. First, sleep homeostasis grows more slowly during the day, allowing teens to remain awake longer without catastrophic levels of sleepiness (5). Second, the circadian timing system undergoes a profound phase delay during puberty (6). This delay puts adolescents at odds with the rhythms of everyday life, creating what the field has dubbed *social jetlag* (7). These two biological shifts are conserved across mammalian phylogeny and thus silence the critique that a teen is simply lazy. When these bioregulatory changes come head-to-head with the innumerable forces placed on teens (early school start times, bussing schedules, athletic obligations, evening homework, even developmentally appropriate needs for socialisation) sleep is measurably and consequently deteriorated.

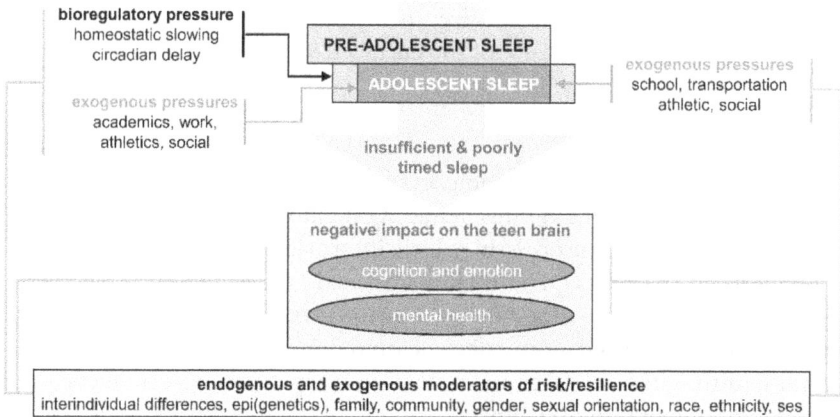

FIGURE 7.1   The 'Perfect Storm' Model of Insufficient Adolescent Sleep Recapitulated. Schematic for understanding insufficient sleep during adolescence: a biological consequence made worse through societal pressures. Top: A simplification of the perfect storm model (3, 4). The bioregulatory changes that occur during puberty: a slowing of sleep homeostatic pressure across wakefulness (Process S) and a delayed circadian timing system (Process C) alter preadolescent sleep by exerting endogenous pressures, moving the ideal sleep opportunity and circadian phase later during adolescence. However, exogenous pressures (e.g., evening demands and school start times) further erode sleep leading to chronically insufficient and poorly timed sleep. Bottom: Consequences of sleep loss and misalignment in adolescents lead to impoverished brain health: cognition, emotion and mental health. Finally – the ultimate quality of life afforded a sleepy teen may be moderated by biological as well as structural factors including economic status and diversity factors across gender, sexual, racial and ethnicity identities.

Groups such as the *Start School Later* movement have pushed to alter school schedules for teens. In the US, California recently became the largest jurisdiction to enact legislation demanding a later (8:30 a.m. at earliest) start time for secondary schools (8). The impact of this change for both teen sleep and its functions remains to be seen. Below we review the science of how insufficient sleep, when left unchecked, can alter brain function and cognitive health, along with societal factors that may impact insufficient sleep in adolescents.

Sleep loss and adolescent cognitive performance

Experimental studies have assessed the impact of sleep restriction on various cognitive processes in adolescents, including memory consolidation (9, 10) and encoding (11), vigilance (12, 13), executive function (13, 14) and abstract thinking (15), as well as on general mood and fatigue. Consensus (16, 17) holds that for adolescents, the effect of sleep restriction may depend both on the cognitive domain and the nature of the curtailed sleep.

Learning and memory

A wealth of data underscores the importance of sleep for learning and memory. Critical for understanding this literature, memory is not one unitary phenomenon, but rather a process in which information, skills or associations are first learned (*encoding*), and then maintained (*consolidation*) to allow for subsequent long-term *recall* and *integration* into knowledge (18). Thus, when considering how reduced sleep affects learning and memory in adolescence, one must distinguish whether reductions in sleep have more influence on some aspects of memory over others.

New learning/encoding

For adolescents, even partial sleep loss can significantly affect new learning and encoding. In one study, Cousins et al. (11) had 15- to 18-year-olds sleep for either five or nine hours per night across five nights, then complete a picture learning task after the 5th night. Following three nights of recovery sleep (nine hours/night), the adolescents who had slept for five hours per night before learning recalled fewer pictures than the nine-hour group. Notably, adolescents' recall was not associated with vigilance or alertness on the day of encoding (similar to prior work at this age; e.g., (19)), suggesting an impact of sleep restriction on encoding ability independent of fatigue or inattentiveness. In a follow-up study (20), 15- to 18-year-olds underwent the same nine- or five-hour-per-night sleep protocol for five nights, but learned novel facts about arthropods (simulating academic content) following the 4th night. Recall tests occurring 30 minutes, three days and 42 days after learning indicated persistent effects of prior sleep restriction on memory, with sleep restricted adolescents recalling 26%, 34% and 65% less material at each respective time-point than those who encoded when rested. Together, these findings indicate that partial sleep restriction can impact adolescents' ability to learn, with long-term consequences for retention.

Memory consolidation and recall

While partial sleep loss negatively impacts adolescents' ability to encode new information, what happens if an adolescent is well-rested during learning but then loses sleep afterward? Contrary to encoding, memory consolidation – the long-term strengthening and storage of previously encoded information – may be less affected by sleep restriction, so long as adolescents are well-rested during learning. Kopasz and colleagues (21) conducted an experiment in which 14- to 16-year-olds engaged in multiple declarative learning tasks (e.g., memory for locations, events, words) after a nine-hour overnight sleep opportunity. Adolescents were then sleep restricted to four hours of overnight sleep (3 a.m.–7 a.m.), followed by a recovery night of nine hours and a recall test. Compared to when adolescents were allotted nine hours, one night of four hours of sleep did not significantly decrease memory recall, suggesting that memory consolidation at this age may be robust, at least in the short term, to partial sleep restriction. In another study, Voderholzer et al. (10) restricted 14- to 16-year-olds across four nights under one of five protocols (9, 8, 7, 6 or 5 hours of time in bed (TIB)). Before restriction, all had a nine-hour sleep opportunity and engaged in two memory tasks (word pairs and procedural learning), with recall assessed twice: after two nights of post-restriction recovery sleep and one month later. At both post-test intervals, memory did not differ across groups for either task, indicating no dose-dependency for either word recall or procedural learning. Furthermore, polysomnography on the last night of sleep restriction indicated that the amount of slow-wave sleep was preserved at the expense of other sleep stages in all protocols, perhaps accounting for comparable memory performance across groups.

More recently, Leong et al. (22) found a similar null effect of longer-term sleep restriction on prospective memory; i.e., memory for actions that must be completed later (for example, remembering to buy items at the grocery store, or to relay a message to a friend). After five nights of either five hours of TIB or nine hours of TIB (both preceded by a week of nine hours TIB), 15- to 18-year-olds across protocols performed comparably (and poorly) on a prospective memory task requiring them to press a certain button in response to specific stimuli. Although the task itself may have been too difficult to detect beneficial effects of longer sleep, Leong et al.'s (22) findings in combination with others imply that if teens have slept well before learning, subsequent short sleep (here, 4–5 hours) may not significantly impair retention.

Although the amount of information adolescents consolidate may not be decreased by partial sleep loss, sleep may still impact which specific information is prioritised for consolidation over others (23). In one study, 15- to 19-year-old adolescents were tasked with memorising a short passage, with specific sections highlighted as 'important' and associated with a reward (9). On Day 3 of a rested baseline (9 hours/night), adolescents were shown the passage for seven minutes and were tested immediately and after one week of either restricted (five hours/night) or control sleep (nine hours/night). While both groups remembered the highlighted sections better than the non-highlighted ones, the mnemonic benefit became even more pronounced at the one-week test only in the control group, suggesting that restricted sleep may impair the adolescent brain's ability to discriminate which information is most critical to retain. Similarly, another study found that partial sleep loss before encoding (five hours/night for seven nights) promotes greater inaccurate/false narratives in eyewitness memories (24). These new lines of research tentatively indicate that sleep loss may affect the adaptiveness and accuracy of adolescent memory consolidation, even if the overall amount of information consolidated is equivalent.

Summary

Recent experiments indicate that partial sleep loss over multiple nights causally disrupts new learning and memory formation in adolescents. Though consolidation of prior memories can still occur, sleep loss can shift the prioritisation of some memories over others for consolidation. Although restriction to 5–6.5 hours of sleep per night (the range used for most recent studies) may seem severe, it is not unreasonable given that <6.5 hours of sleep is common for over a quarter of adolescents (25). Nevertheless, future research should address the effects of more nuanced levels of sleep loss for new learning, as some studies of consolidation have done.

Attention, executive functioning and self-regulation

Sleep loss's impact on adolescent attention and executive functioning is highly relevant for daily living. Executive functioning encompasses processes such as self-regulation, working memory and flexibility, whereas attention involves balancing exploitation and exploration of competing stimuli and information sources (26, 27). For adolescents, attention and executive functions are critical for newly learned activities such as driving. With motor vehicle accidents identified as the second leading cause of death for adolescents in the US (28, 29), it is crucial to study how sleep loss impacts capacities critical to avoid such events.

Attention and vigilance

The impact of sleep loss on vigilance and attention has been long studied in adults. Comparing the effects of staying up late to those of alcohol consumption, researchers in the 1990s found that 18–19 hours of sustained wakefulness produced vigilance impairments comparable to those found among adults with a blood alcohol concentration of 0.05% (30, 31). Furthermore, when sober participants were kept awake for 24 hours, performance on an unpredictable tracking task (using a joystick to follow a constantly moving target) was as poor as that of individuals with a blood alcohol content of 0.10% (above the US legal limit).

Investigating such effects in adolescents is a more recent endeavour, with studies using both laboratory tasks and ecologically relevant measures to substantiate that sleep loss impairs adolescent attention and vigilance. One popular laboratory measure of sustained attention is the psychomotor vigilance task (PVT), in which individuals must continuously attend and respond to a visual stimulus appearing at irregular randomised intervals. When adolescents' sleep is restricted across days, PVT performance deteriorates, with adolescents showing more failures to respond ('lapses'; (32, 33)) and fewer timely responses relative to premature or late reactions (34). Vigilance in young adolescents (~10–11 years) appears particularly vulnerable, as PVT performance decline has been observed even with relatively minor sleep loss (i.e., 10-hour TIB versus 8.5-hour TIB; (34)). In contrast, Campbell et al. (34) found that performance in older adolescents (~15–16 years) did not differ between four days of ten-hour TIB and 8.5-hour TIB. Nonetheless, sleep reduction to a seven-hour opportunity produced clear PVT deficits in both older and younger adolescents.

Agostini et al. (33) had previously reported that severe sleep restriction (five hours TIB/night for five nights) produced PVT deficits in 15- to 17-year-olds when compared to baseline, and that performance was not restored even after two days of recovery sleep. Subsequently, Short et al. (32) modelled older adolescents' sleep need and PVT performance using five-day restriction protocols of five-hour and 7.5-hour TIB/night, finding dose-response effects of both sleep opportunity and number of restriction nights. Both five hours/night and 7.5 hours/night resulted in PVT deficits among 15- to 17-year-olds, but deficits were apparent more quickly in the five hours/night condition (i.e., as early as the third day) than in the 7.5 hours/night condition, in which at least five days were required to demonstrate a significant deficit. Short's modelling further indicated that adolescents should obtain 9.35 hours of sleep per night on average for optimal attention, despite prior findings suggesting short-term resilience to 8.5 hours/night sleep restriction at the same age (34). Overall, these studies provide strong evidence that continued sleep loss impairs vigilance in adolescents, with effects possibly moderated by both the specific amount of sleep lost per day and total sleep loss accumulation across days.

They highlight that adolescence is not monolithic, with key differences arising in young (10–11 years) and older (15–18 years) teens. Understanding the trajectory of these effects is critical to both tease apart mechanisms of cognitive development and bolster the ecological relevance of such data.

Though adolescent sleep loss produces vigilance deficits in tightly controlled experiments, such designs lend one to ask whether these deficits would persist in more ecologically valid settings. One recent study in Israel measured the sleep patterns of adolescents (grades 7–12, mean age 16) using wrist-worn actigraphy on school nights and weekends (35). After showing that adolescents' sleep was indeed shorter on school nights, they then demonstrated comparatively slower reaction times and more lapses on the PVT on subsequent days. Though these results could be confounded by other factors such as increased stress on weekdays relative to weekends, they are in line with the laboratory-based studies above. In another study, researchers experimentally manipulated sleep via two 5-night sleep conditions (short sleep: 6.5 hours/night TIB; long sleep: ten hours/night) but had participants (aged 16–18 years) sleep at home rather than in the laboratory (36). Furthermore, rather than using the PVT, participants completed a simulated driving task to capture a more naturalistic context. Adolescents showed less reliable vehicle control in a rural driving scenario (e.g., lateral drift within their lane) when they were sleep restricted. Together, these studies reinforce that: a) adolescents experience behaviourally impactful sleep curtailment naturally in their everyday lives; and b) the effects of adolescent sleep loss extend to ecologically valid tasks with implications for safety and well-being.

One tentative silver lining in an otherwise sobering account of sleep loss's effects on adolescent attention is that these effects can be alleviated to some extent, at least in older adolescents, by daytime naps. Using a 2-week sleep protocol (with the first 5 days preceded and followed by two nights of extended 'weekend' recovery sleep) and the PVT as an outcome, Lo and colleagues (37) had 15- to 19-year-old adolescents follow nocturnal sleep restriction either with or without daytime naps. For adolescents whose nocturnal sleep was restricted to 6.5 hours TIB, a 1.5-hr daily nap opportunity in the early afternoon allowed them to maintain the same attentiveness levels as after baseline sleep. When adolescents' nocturnal sleep was restricted to five hours, napping for 1–1.5 hours similarly alleviated PVT deficits relative to not napping, but adolescents remained impaired. Thus, these findings support a benefit of naps for older teenagers when a full night of sleep is not possible, though the palliative effects of a nap may depend on the severity of overnight sleep loss. Further work must examine these possibilities in younger adolescents.

Working memory

Working memory is often defined as the ability to hold recent information in mind to inform ongoing behaviour. Under this definition, results are currently mixed regarding how sleep loss impacts adolescent working memory, complicated by inconsistencies in the age ranges studied, the working memory paradigm used and the degrees of sleep loss implemented at different ages. Specifically, recent working memory studies have mostly restricted sleep either to five hours (12, 13) or 6–6.5 hours TIB per night across 5–7 consecutive nights (14, 38–41). In such studies, the most common test employed is the 'N-back' task, in which participants are shown symbols one at a time and must remember whether the current symbol matches that shown a certain number (N) of trials previously. With sleep restricted to five hours per night, adolescents have shown significant performance deficits as early as after night three of restriction (12, 13), whereas adolescents who are only 'mildly' sleep restricted to 6.5 hours/night maintain baseline performance through at least five consecutive nights (38, 39). In contrast, with a visuospatial token task (in which adolescents were required to remember locations of tokens so as not to revisit those locations), Kiriş (14) recently reported that 18- to 19-year-olds showed deteriorated performance after only four nights even when sleep restriction was relatively mild. Similarly, Jiang and colleagues (41) found that 13- to 16-year-old adolescents were slower to respond on serial subtraction and verbal working memory tasks after six-hour TIB/night across five nights, despite no change in accuracy. Thus, sleep restriction as mild as 6- to 6.5-hour TIB/night may produce behavioural changes for some paradigms, though which task demands are required to observe these deficits requires further clarification.

Whether mild sleep loss impairs working memory may also depend on the distribution and timing of sleep bouts. Lo et al. (40) found that when 15- to 19-year-old adolescents were sleep restricted to 6.5 hours daily, they performed better on one- and three-back tasks if their sleep was split into two unequal bouts (five hours of overnight sleep and a 1.5-hour daytime nap) compared to 6.5 hours of overnight sleep. This finding is encouraging for adolescents who cannot feasibly obtain the recommended amount of sleep overnight. Nevertheless, this study is limited by its relatively brief protocol (five consecutive restriction nights and three more after recovery sleep), leaving it unclear whether the beneficial effect of napping would persist across longer-term sleep restriction.

Apart from behavioural differences in working memory, researchers employing fMRI have found that adolescent sleep schedules of 6.5-hour TIB/night will produce differences in neural responses to the N-back task. Beebe et al. (39) reported that after 13.9- to 16.9-year-old adolescents were sleep restricted to 6.5-hour TIB/night for five nights, neural activation in task-positive regions (regions typically active during the task) and neural suppression in task-negative regions (regions typically inactive during the task) were both enhanced during a two-back task when compared to extended sleep (ten-hour TIB). Behavioural performance was similar across the two sleep schedules. Alsameen et al. (38) subsequently examined whether manipulating N-back task difficulty (i.e., from zero-back to three-back) would produce differences in both behavioural and neural outcomes in sleep restricted 14- to 16.9-year-olds. However, after five nights of 6.5-hour TIB, adolescents did not show any behavioural deficits at any difficulty level. Nevertheless, sleep loss impacted adolescents' brain activity, with some brain areas (e.g., medial prefrontal) exhibiting compensatory activation or suppression while others showed a weakening of compensatory activity, particularly during the most difficult task level. Together, these studies indicate that even in the absence of behavioural effects, sleep loss may still result in detectable changes to neural systems underlying working memory. It remains to be determined whether inconsistencies in working memory deficits systematically result from the degree of sleep loss per se or to differences in participants' ages between studies.

Self-regulation, mood and mental health

Bidirectional associations between poor sleep and self-regulation difficulties have been reported in both adults and adolescents, with irregular and insufficient sleep linked to higher rates of depression, anxiety, suicidal ideation and risky decision-making (42, 43) (also see (25, 44) for reviews). In large US samples, the highest odds of serious suicide attempts were observed in adolescents reporting the shortest sleep (≤4 hours per night; (45)). Excessive oversleeping (≥ten hours per night) has also been related to teen suicidality (45), and the effects of sleep on mood may be moderated by demographic factors such as biological sex, gender identity, race and ethnicity (42). Overall, disrupted sleep in both extremes – too much and too little – have been forwarded as potential causes and consequences of altered teen mental health.

Experimentally, many adolescent studies have taken advantage of self- and parent-reported questionnaires to probe the causal effect of sleep loss on mood and self-regulation. Unsurprisingly, most studies have found a negative effect of sleep loss on adolescent mood; experimentally sleep-deprived teens and their caregivers report decreases in teens' positive mood and vigour with increases in confusion, fatigue, hostility and anxiety (13, 46–48). Recently, Booth et al. (47) indicated that while 15- to 17-year-olds' self-ratings of depression, anger and happiness were relatively unchanged by five nights of sleep restriction to 7.5 hours per night, sleep restriction to five hours per night significantly increased depression and anger and decreased happiness. Among 14- to 17-year-olds however, Baum et al. (46) reported that milder sleep restriction – 6.5 hours TIB per night for five nights – also increased self-reported anger, anxiety and confusion and parent-corroborated irritability, suggesting that this intermediate level of sleep restriction may be enough to trigger emotional dysregulation in some adolescents. These findings specify that for older teenagers, sleep curtailment to 6.5 hours or less may result in significant emotional dysregulation.

One obvious limitation of these questionnaire-based studies is participants' own expectations in self-reporting. If adolescents know they will be losing sleep, they likely also expect to be more irritable or 'moody'. While most studies attempt to account for expectancy effects, either by embedding the mood ratings into other control questionnaires (46) or by tightly regulating participants' sleep/wake environmental cues (47), it is uncertain just how successful such methods are at reducing adolescents' expectations. More recent studies have thus supplemented questionnaires with task-based emotion and mood measures, such as pupil response (48) and behavioural reactivity to emotion-inducing stimuli or peer conflicts (48, 49). Overall, these alternative measures corroborate adolescents' subjective reports, though the studies employing them have been limited in their range of sleep restriction. Both McMakin et al. (48) and Reddy et al. (49) used acute restriction paradigms of four hours TIB for one or two nights. Further work should extend these tasks to the sorts of milder, longer-term sleep restriction protocols described above.

Adolescent sleep-related mood disturbances may also be partly eased by daytime napping. In Lo et al.'s (40) study, 15- to 19-year-olds who were sleep restricted to 6.5 hours per day sleep schedules reported higher positive mood when their sleep was split into a 1.5-hour daytime nap and five hours overnight compared to 6.5 hours of overnight sleep. When participants were permitted to sleep for eight hours, mood ratings remained stable regardless of whether these eight hours were split or continuous. Thus, as with other domains, adolescent mood and regulation may benefit from napping, and a split sleep schedule may be a helpful alternative for busy teens when eight hours of continuous overnight sleep is not possible.

Summary

Sleep loss undeniably has the potential to negatively affect adolescents' attention and executive functioning, including working memory and self-regulation. The current literature suggests that attention and self-regulation are perhaps the most vulnerable to even mild sleep loss, whereas the effects of sleep loss on working memory may be more nuanced and dependent on the specific task, degree of sleep loss or age range studied. However, because individual studies often differ in the age ranges assessed and the degree/duration of sleep restriction implemented, it is difficult to draw developmental conclusions at this time. Furthermore, along with sleep relating to mental health via mood, it should be noted that there are studies analysing the exacerbating effects of sleep loss in other psychiatric and neurodevelopmental conditions (e.g., ADHD) that are outside the realms of this chapter.

Societal factors influencing insufficient sleep and its consequences: opportunities for buffering?

In the previous sections, we have identified primary social and biological factors contributing to sleep loss across adolescence, as well as their adverse effects on cognitive and emotional function. In light of what we know about adolescents' unique propensity for sleep loss and its effects on functioning, one final topic of critical importance concerns identifying malleable societal structures that may moderate teen sleep.

School start times

In response to our knowledge of the physiological and social changes compelling adolescents to stay up later in the evenings, some US districts have implemented delayed school start times, with California enacting the first state-wide law requiring high schools to start no earlier than 8:30 a.m. (8). Both school-wide and multisite studies investigating the effects of such policies have reported promising results, with later start times associated with increases in adolescent school attendance (50), longer sleep (51, 52), higher grades (53) and better mental health (44, 50, 53) across both suburban and urban school districts. For adolescents old enough to drive, delaying school start times has led to reductions in drowsy driving and adolescent motor vehicle accidents (28, 53). Finally, changing school start times may aid students with chronotypes that are even later than those predicted by typical adolescent development (see (54) for a discussion of chronotype-specific personalised school start times).

Embracing naps

Although delaying school has already gained momentum in the US and internationally, we must also recognise that logistical and resource-related barriers often prevent this intervention from being feasible. In these cases, daytime napping may be another option to help ameliorate adolescent sleep loss if timed to not substantially influence night-time sleep pressure. Given the nap benefits observed experimentally for memory, attention, executive functions, and self-regulation (37, 40, 55, 56), it is clear that napping benefits adolescent functioning particularly when sleep restricted. However, questions that remain include the duration of nap opportunity that should be provided in schools, and what the optimal timing of the nap period should be relative to varying class schedules.

Addressing socioeconomic and racial/cultural inequities

Alongside implementing sleep interventions directly in schools, addressing social and environmental inequities corresponding to differences in socioeconomic status (SES), race and ethnicity may also help to improve sleep for adolescents. Pérez Ortega (57) recently reviewed factors underlying racial and ethnic disparities in sleep quantity and quality for US adults, pointing to differences in job opportunities/shift work, acculturation stress, racial discrimination, exposure to increased air and light pollution and a lack of culturally responsive education about sleep hygiene on the part of doctors and scientists as contributors to these disparities. Among adolescents, research has mainly focused on family and neighbourhood SES as contributors to disrupted teen sleep (58, 59), with fewer studies focused on the role of external environmental noise, light (outside of personal electronics use) or air pollutants. However, one large-scale study in China indicated that greater exposure to small particulate air pollution predicted greater sleep problems in children aged 2–17 (60). Neighbourhood status in both the US and abroad predicts shorter sleep duration in children and adolescents (ages 0–18; (61)), whereas familial economic instability and caregiver stress have been more closely associated with variability in teen sleep rather than total sleep time (62). Familial SES has also been shown to moderate associations between adolescent sleep and emotional regulation, with adolescents from lower SES backgrounds demonstrating the greatest link between good sleep and emotional function (63).

From a cultural perspective, even fewer studies have evaluated how migration, acculturation or discrimination might affect sleep in adolescence. However, one study evaluating predictors of sleep loss in Mexican-American adolescents (US grade 7 and above) indicated that higher parent acculturation, income and education were paradoxically (along with neighbourhood crime) associated with poorer adolescent sleep health, whereas greater family unity and lower acculturation was related to better sleep health (64). Such results illuminate the importance of an intersectional perspective. Aside from this study, more recent work has focused on how sleep may serve as a protective factor against acculturation stress in high school students migrating to the US from various countries, finding that longer sleep durations predicted a more pronounced decrease in stress and greater cultural adaptation across the first year of immigration (65). Holistically, these studies emphasise a need to consider bidirectional relations between culture-specific variables and sleep health in adolescence, and to evaluate how societal inequities beyond SES may impact sleep and its effect on waking function.

Summary

To improve adolescent sleep, possible interventions include delaying school start times, encouraging daytime naps and addressing broader social inequities contributing to teen sleep loss. With the current forward momentum of policies implementing delayed start times, evidence in favour of this practice is rapidly accumulating, with beneficial implications for students and minimal negative effects. Even so, personalised interventions may need to account for students' individual identities and implement additional strategies to promote sufficient sleep.

Conclusion and urgent research needs

The current chapter reviewed literature indicating the impact of insufficient sleep on developing adolescents' cognition, emotion and mental health, while also pointing to areas of possible intervention. Many of the studies presented provide evidence that parallels the negative impact of sleep loss in adults. However, the root causes of insufficient sleep in adolescents are unique yet well-known scientifically. Insufficient sleep in teens – even of only 1–2 hours a night – when occurring on a near-nightly basis can have, as it does in adults, a profound impact on waking function.

Attention, learning, memory and emotional regulation are all compromised by insufficient sleep with no permanent countermeasure present (even from napping). Unlike in adults, however, teens often lack the agency to change their schedules to better accommodate sleep. The impact of insufficient sleep on cognitive and affective brain function imposed on adolescents coincides with a critical window for psychosocial development, as teens navigate an ever-changing landscape of motivations, opportunities, risks and learning critical for their future quality of life (1). The emergence of mental illness in the second decade provides only an additional lens from which to consider the growing cost of sleep loss for young people. Finally, societal factors including but not limited to economic status, school timing and racial and ethnic health disparities can not only increase the pressures placed on the sleep regulatory system but also moderate or mediate the impact that compromised sleep can have during development. Taken together, the evidence presented here and elsewhere places our field in a unique position. We believe there are at least five distinct needs at this juncture which we highlight below to close this chapter:

First, we need truly developmental experiments. While there is utility in documenting adult-like effects in cross-sectional samples of adolescents, we must also turn to what makes adolescents different than adults. Particularly when experiments include mechanistic neurobiology (be it fMRI of the sleep-deprived brain, EEG of recovery sleep, or hormonal assessments of circadian factors such as melatonin), future work needs to be implicitly placed within a developmental framework. Through such a lens we can identify not only *if* adolescents differ from adults in their susceptibility to sleep loss but also *how*, why and when. Understanding how neurodevelopment interacts with sleep need and is differentially moderated by sleep loss is paramount.

Second, our studies need to progress towards increased ecological validity. Our protocols must move beyond sleep loss manipulations. Total sleep deprivation is an effective manipulation yet not the reality of sleep loss for the vast majority of individuals (save an unfortunate *all nighter*). Chronic sleep restriction moves the needle one step closer towards ecological validity, yet they too lack the true nuance of adolescent life: the co-occurring circadian social jetlag.

The quest for ecological validity extends to our assessments. With much of the public discourse on adolescent sleep need focused on schools, the use of tests directly related to educational achievement and social health – rather than simply our laboratory tests of neurobehavioral function – will close the gap between the literature and the quotidian experience of under-slept teens.

Third, most studies in adolescents focus on whether an effect of sleep loss is present rather than for whom the effect is greatest. We know from data in adults that phenotypic resilience and vulnerability to sleep loss emerges for specific outcomes (e.g., (66)). Moderating factors such as the trajectories of neurodevelopment and the emergence of atypical development (e.g., ADHD, autism) may provide additional pressures determining any one child's individual susceptibility to sleep loss (67). While policy initiatives such as that from California in the US pursue one-size-fits-all increases in sleep opportunity for youth, more phenotypic work would support person-centred solutions.

Fourth, a need for intersectional samples is clear. Interacting factors such as economic, race and ethnic minority status play a large role in risk for impaired physical or mental health in teens in a way that may dovetail back to moderate sleep loss effects. Only by intentionally expanding our studies to underserved teens – and by including explicit tests of moderating influences – can we begin to understand the universality of these effects.

Finally, while this chapter has been principally concerned with sleep loss, future studies must turn to what can be done to buffer resilience. While adolescent biology is a given, and structural factors such as school start times are slow to change, it is incumbent on our work to identify ways to increase sleep at the margin. We can begin to understand the trade-off between how long a teen sleeps and when they sleep to provide better recommendations for teens facing an impossible problem. The short-term and long-term benefits of countermeasures such as naps and even caffeine must be thoroughly examined in the lens of brain-based cognitive and affective health.

More than simply an empirical neuroscientific problem, sleep loss during adolescence is a pressing public health need. Future studies are encouraged to keep that translatability in mind when considering their experimental designs, populations, assessments and ultimate scalability.

References

1 Dahl RE. Adolescent brain development: a period of vulnerabilities and opportunities. Keynote address. Ann N Y Acad Sci. 2004;1021(1):1–22.
2 2004 sleep in America poll: summary of findings. Vol. 4, National Sleep Foundation. 2004. p. 1–58. Available from: www.thensf.org/wp-content/uploads/2021/03/2004-SIA-Findings.pdf
3 Carskadon MA. Sleep in adolescents: the perfect storm. Pediatr Clin North Am. 2011;58(3):637–47.

4 Crowley SJ, Wolfson AR, Tarokh L, Carskadon MA. An update on adolescent sleep: new evidence informing the perfect storm model. J Adolesc. 2018;67:55–65.

5 Jenni OG, Achermann P, Carskadon MA. Homeostatic sleep regulation in adolescents. Sleep. 2005;28(11):1446–54.

6 Roenneberg T, Kuehnle T, Pramstaller PP, Ricken J, Havel M, Guth A, et al. A marker for the end of adolescence. Curr Biol. 2004;14(24):R1038–9.

7 Wittmann M, Dinich J, Merrow M, Roenneberg T. Social jetlag: misalignment of biological and social time. Chronobiol Int. 2006:497–509.

8 Ziporyn TD, Owens JA, Wahlstrom KL, Wolfson AR, Troxel WM, Saletin JM, et al. Adolescent sleep health and school start times: setting the research agenda for California and beyond. A research summit summary. Sleep Heal. 2022;8(1):11–22.

9 Lo JC, Bennion KA, Chee MWL. Sleep restriction can attenuate prioritization benefits on declarative memory consolidation. J Sleep Res. 2016;25(6):664–72.

10 Voderholzer U, Piosczyk H, Holz J, Landmann N, Feige B, Loessl B, et al. Sleep restriction over several days does not affect long-term recall of declarative and procedural memories in adolescents. Sleep Med. 2011;12(2):170–8.

11 Cousins JN, Sasmita K, Chee MWL. Memory encoding is impaired after multiple nights of partial sleep restriction. J Sleep Res. 2018;27(1):138–45.

12 Lo JC, Lee SM, Teo LM, Lim J, Gooley JJ, Chee MWL. Neurobehavioral impact of successive cycles of sleep restriction with and without naps in adolescents. Sleep. 2017;40(2).

13 Lo JC, Ong JL, Leong RLF, Gooley JJ, Chee MWL. Cognitive performance, sleepiness, and mood in partially sleep deprived adolescents: the need for sleep study. Sleep. 2016;39(3):687–98.

14 Kiriş N. Effects of partial sleep deprivation on prefrontal cognitive functions in adolescents. Sleep Biol Rhythms. 2022;20(4):499–508.

15 Randazzo AC, Muehlbach MJ, Schweitzer PK, Waish JK, Walsh JK. Cognitive function following acute sleep restriction in children ages 10–14. Sleep. 1998;21(8):861–8.

16 Kopasz M, Loessl B, Hornyak M, Riemann D, Nissen C, Piosczyk H, et al. Sleep and memory in healthy children and adolescents – a critical review. Sleep Med Rev. 2010;14(3):167–77.

17 Short MA, Chee MWL. Adolescent sleep restriction effects on cognition and mood. In: Progress in Brain Research; 2019. https://doi.org/10.1016/bs.pbr.2019.02.008

18 Diekelmann S, Wilhelm I, Born J. The whats and whens of sleep-dependent memory consolidation. Sleep Med Rev. 2009;13(5):309–21.

19 Beebe DW, Field J, Milller MM, Miller LE, LeBlond E. Impact of multi-night experimentally induced short sleep on adolescent performance in a simulated classroom. Sleep. 2017;40(2).

20 Cousins JN, Wong KF, Chee MWL. Multi-night sleep restriction impairs long-term retention of factual knowledge in adolescents. J Adolesc Heal. 2019;65(4):549–57.

21 Kopasz M, Loessl B, Valerius G, Koenig E, Matthaeas N, Hornyak M, et al. No persisting effect of partial sleep curtailment on cognitive performance and declarative memory recall in adolescents. J Sleep Res. 2010;19(1):71–9.

22 Leong RLF, Koh SYJ, Tandi J, Chee MWL, Lo JC. Multiple nights of partial sleep deprivation do not affect prospective remembering at long delays. Sleep Med. 2018;44:19–23.

23 Saletin JM, Walker MP. Nocturnal mnemonics: sleep and hippocampal memory processing. Front Neurol. 2012;3:1–12.

24 Lo JC, Chong PLH, Ganesan S, Leong RLF, Chee MWL. Sleep deprivation increases formation of false memory. J Sleep Res. 2016;25(6):673–82.

25 Moore M, Meltzer LJ. The sleepy adolescent: causes and consequences of sleepiness in teens. Paediatr Respir Rev. 2008;9(2):114–21.

26 Aston-Jones G, Cohen JD. An integrative theory of locus coeruleus-norepinephrine function: adaptive gain and optimal performance. Annu Rev Neurosci. 2005;28:403–50.

27 Krauzlis RJ, Bollimunta A, Arcizet F, Wang L. Attention as an effect not a cause. Trends Cogn Sci. 2014;1–8.

28 Meltzer LJ, Plog AE, Swenka D, Reeves D, Wahlstrom KL. Drowsy driving and teen motor vehicle crashes: impact of changing school start times. J Adolesc. 2022;94(5):800–5.

29 Yellman MA, Bryan L, Sauber-Schatz EK, Brener N. Transportation risk behaviors among U.S. high school students – youth risk behavior survey, United States, 2019. MMWR Suppl. 2020;69(1):77–83.

30 Dawson D, Reid K. Fatigue, alcohol and performance impairment. Nature. 1997;388(6639):235.

31 Lamond N, Dawson D. Quantifying the performance impairment associated with fatigue. J Sleep Res. 1999;8(4):255–62.

32 Short MA, Weber N, Reynolds C, Coussens S, Carskadon MA. Estimating adolescent sleep need using dose-response modeling. Sleep. 2018;41(4):1–14.

33 Agostini A, Carskadon MA, Dorrian J, Coussens S, Short MA. An experimental study of adolescent sleep restriction during a simulated school week: changes in phase, sleep staging, performance and sleepiness. J Sleep Res. 2017;26(2):227–35.

34 Campbell IG, Van Dongen HPA, Gainer M, Karmouta E, Feinberg I. Differential and interacting effects of age and sleep restriction on daytime sleepiness and vigilance in adolescence: a longitudinal study. Sleep. 2018;41(12):1–8.

35 Orna T, Efrat B. Sleep loss, daytime sleepiness, and neurobehavioral performance among adolescents: a field study. Clocks Sleep. 2022;4(1):160–71.

36 Garner AA, Miller MM, Field J, Noe O, Smith Z, Beebe DW. Impact of experimentally manipulated sleep on adolescent simulated driving. Sleep Med. 2015;16(6):796–9.

37 Lo JCY, Koa TB, Ong JL, Gooley JJ, Chee MWL. Staying vigilant during recurrent sleep restriction: dose-response effects of time-in-bed and benefits of daytime napping. Sleep. 2022;45(4):1–9.

38 Alsameen M, DiFrancesco MW, Drummond SPA, Franzen PL, Beebe DW. Neuronal activation and performance changes in working memory induced by chronic sleep restriction in adolescents. J Sleep Res. 2021;30(5):1–11.

39 Beebe DW, DiFrancesco MW, Tlustos SJ, McNally KA, Holland SK. Preliminary fMRI findings in experimentally sleep-restricted adolescents engaged in a working memory task. Behav Brain Funct. 2009;5:1–7.

40 Lo JC, Leong RLF, Ng ASC, Jamaluddin SA, Ong JL, Ghorbani S, et al. Cognitive effects of split and continuous sleep schedules in adolescents differ according to total sleep opportunity. Sleep. 2020;43(12):1–11 [cited 2022 Dec 4].

41 Jiang F, Vandyke RD, Zhang J, Li F, Gozal D, Shen X. Effect of chronic sleep restriction on sleepiness and working memory in adolescents and young adults. J Clin Exp Neuropsychol. 2011;33(8):892–900.

42 Winsler A, Deutsch A, Vorona RD, Payne PA, Szklo-Coxe M. Sleepless in Fairfax: the difference one more hour of sleep can make for teen hopelessness, suicidal ideation, and substance use. J Youth Adolesc. 2015;44(2):362–78.

43 Zhang J, Paksarian D, Lamers F, Hickie IB, He J, Merikangas KR. Sleep patterns and mental health correlates in US adolescents. J Pediatr. 2017;182:137–43.

44 Owens J, Au R, Carskadon M, Millman R, Wolfson A, Braverman PK, et al. Insufficient sleep in adolescents and young adults: an update on causes and consequences. Pediatrics. 2014;134(3):e921–32.

45 Fitzgerald CT, Messias E, Buysse DJ. Teen sleep and suicidality: results from the youth risk behavior surveys of 2007 and 2009. J Clin Sleep Med. 2011;7(4):351–6.

46 Baum KT, Desai A, Field J, Miller LE, Rausch J, Beebe DW. Sleep restriction worsens mood and emotion regulation in adolescents. J Child Psychol Psychiatry Allied Discip. 2014;55(2):180–90.

47 Booth SA, Carskadon MA, Young R, Short MA. Sleep duration and mood in adolescents: an experimental study. Sleep. 2021;44(5).

48 McMakin DL, Dahl RE, Buysse DJ, Cousins JC, Forbes EE, Silk JS, et al. The impact of experimental sleep restriction on affective functioning in social and nonsocial contexts among adolescents. J Child Psychol Psychiatry Allied Discip. 2016;57(9):1027–37.

49 Reddy R, Palmer CA, Jackson C, Farris SG, Alfano CA. Impact of sleep restriction versus idealized sleep on emotional experience, reactivity and regulation in healthy adolescents. J Sleep Res. 2017;26(4):516–25.

50 Wahlstrom K. Changing times: findings from the first longitudinal study of later high school start times. NASSP Bull. 2002;86(633):3–21.

51 Nahmod NG, Lee S, Master L, Chang AM, Hale L, Buxton OM. Later high school start times associated with longer actigraphic sleep duration in adolescents. Sleep. 2019;42(2):1–10.

52 Widome R, Berger AT, Iber C, Wahlstrom K, Laska MN, Kilian G, et al. Association of delaying school start time with sleep duration, timing, and quality among adolescents. JAMA Pediatr. 2020;174(7):697–704.

53 Wahlstrom KL. Examining the impact of later high school start times on the health and academic performance of high school students: a multi-site study final report. Cent Appl Res Educ Improv Univ Minnesota Minneapolis, MN, USA; 2014.

54 Goldin AP, Sigman M, Braier G, Golombek DA, Leone MJ. Interplay of chronotype and school timing predicts school performance. Nat Hum Behav. 2020;4(4):387–96.

55 Lemos N, Weissheimer J, Ribeiro S. Naps in school can enhance the duration of declarative memories learned by adolescents. Front Syst Neurosci. 2014;8:103.

56 Leong RLF, Yu N, Ong JL, Ng ASC, Jamaluddin SA, Cousins JN, et al. Memory performance following napping in habitual and non-habitual nappers. Sleep. 2021;44(6):1–11.

57 Pérez Ortega R. Divided we sleep. Science. 2021;374(6567):552–5. Available from: www.science.org/doi/10.1126/science.acx9445

58 Marco CA, Wolfson AR, Sparling M, Azuaje A. Family socioeconomic status and sleep patterns of young adolescents. Behav Sleep Med. 2011;10(1):70–80.

59 Mayne SL, Mitchell JA, Virudachalam S, Fiks AG, Williamson AA. Neighborhood environments and sleep among children and adolescents: a systematic review. Sleep Med Rev. 2021;57(2021):101465.

60 Lawrence WR, Yang M, Zhang C, Liu RQ, Lin S, Wang SQ, et al. Association between long-term exposure to air pollution and sleep disorder in Chinese children: the Seven Northeastern Cities study. Sleep. 2018;41(9):1–10.

61 Tomfohr-Madsen L, Cameron EE, Dhillon A, MacKinnon A, Hernandez L, Madigan S, et al. Neighborhood socioeconomic status and child sleep duration: a systematic review and meta-analysis. Sleep Heal. 2020;6(5):550–62.

62 Schmeer KK, Tarrence J, Browning CR, Calder CA, Ford JL, Boettner B. Family contexts and sleep during adolescence. SSM – Popul Heal. 2019:100320.

63 El-Sheikh M, Shimizu M, Philbrook LE, Erath SA, Buckhalt JA. Sleep and development in adolescence in the context of socioeconomic disadvantage. J Adolesc. 2020;83:1–11.

64 McHale SM, Kim JY, Kan M, Updegraff KA. Sleep in Mexican-American adolescents: social ecological and well-being correlates. J Youth Adolesc. 2011;40(6):666–79.

65 Venta A, Alfano C. Can sleep facilitate adaptation for immigrant high schoolers? Longitudinal relations between sleep duration, acculturative stress, and acculturation. Child Psychiatry Hum Dev. 2021;54(1):147–53.

66 Galli O, Jones CW, Larson O, Basner M, Dinges DF. Predictors of interindividual differences in vulnerability to neurobehavioral consequences of chronic partial sleep restriction. Sleep. 2022;45(1):1–14.
67 Owens J, Gruber R, Brown T, Corkum P, Cortese S, O'Brien L, et al. Future research directions in sleep and ADHD: report of a consensus working group. J Atten Disord. 2013;17(7):550–64.

8

MECHANISTIC ROLE OF SLEEP IN CARDIOVASCULAR AND METABOLIC DISEASES

Elizabeth F. Rasmussen, Suzanne B. Gorovoy and Michael A. Grandner

Introduction

The field of sleep medicine has evolved from the 1970s sleep research field to encompass classification of sleep stages and investigation of sleep disorders largely due to the discovery of electroencephalogram (EEG) patterns that occur during sleep. Since the early 2000s, sleep medicine research has accelerated, and the amount of peer-reviewed sleep journals has more than tripled (1). Aspects of sleep and sleep health including duration, quality, latency, daytime dysfunction, habitual sleep, sleep medication and sleep disturbances have been studied in a wide variety of contexts and ten sleep-wake disorder categories are classified in the Diagnostic and Statistical Manual of Mental Disorders (DSM-5) (2). The role of sleep has been investigated in metabolism, hormone regulation and gene expression (3), and poor sleep has been linked to a variety of disorders including hypertension (4), type-2 diabetes (5), immune function (6), dementia (7), mood disorders and cardiovascular disease (8). This chapter will focus on the mechanistic role of sleep in cardiometabolic health. Through examining sleep-associated physiologic alterations in emotional functioning, decision-making, energy level, physiologic regulation, metabolic rhythms, and immunologic function, molecular and mechanistic pathways by which these alterations impact cardiometabolic health will be explored.

Defining sleep health in the study of cardiometabolic disease: caveats and qualifications

For the purposes of understanding the mechanistic role of sleep in cardiovascular and metabolic diseases, it is important to recognise that the majority of

DOI: 10.4324/9781003296966-8

studies that have explored this have examined single dimensions of sleep at a time. Like dietary/nutritional health, sleep health is a multidimensional concept (9). Most studied is sleep duration. Many studies now exist that suggest that insufficient sleep duration (and, in some cases, excessive sleep duration) is specifically associated with a wide range of adverse health outcomes (10).

Still, even this unidimensional definition of sleep health contains important nuances requiring qualification. For example, sometimes this is based on retrospective self-report (e.g., 'How much sleep do you typically get?'), sometimes this is based on objective, prospective assessment of habitual sleep duration (e.g., several days of assessment with actigraphy), and sometimes this is based on acute modification of sleep, often in a laboratory (e.g., sleep deprivation, partial sleep deprivation, or sleep extension) (11). Although all these model various aspects of sleep sufficiency/insufficiency, they do so in different ways (12, 13). In addition, neither laboratory manipulation of sleep nor single-timepoint assessment of habitual sleep captures the dynamic nature of sleep which can change over time. Even the few studies with more than one timepoint typically assess only two periods, often years apart (14). It is unclear how the dynamics of changes in sleep duration impact outcomes.

In the context of all these issues, there are yet other dimensions of sleep health that are separate from sleep duration (3). For example, timing and regularity of sleep may reflect circadian dysregulation or desynchronisation. Poor sleep quality, separate from duration, may be represented by self-reported dissatisfaction with sleep, sleep fragmentation, interruptions to sleep, suboptimal sleep architecture, sleep continuity disturbances, etc. All of these may be independently associated with health outcomes. Daytime sleepiness may also represent an element of sleep health as either a consequence of residual homeostatic sleep pressure or as a separate process. These dimensions may overlap, cause each other, and represent both cause and consequence of adverse outcomes.

Additionally, sleep disorders may confer cardiometabolic risk separately from their direct influence on sleep parameters described above. For example, insomnia may contribute to health outcomes through more wake time, irregularity and sleep continuity disturbances, but it may also contribute in other ways (15) since insomnia mechanisms often overlap with those of mental health, which may independently confer risk, and the hyperarousal itself may separately contribute as well. Additionally, the intermittent hypoxia of sleep apnea may confer risk through alterations to sleep but it also has independent contributions to cellular function (16).

Many sleep disturbances are also related to social, behavioural and environmental risk factors that independently confer risk (10). For example, poor sleep health in real-world settings may partially reflect socioeconomic status, occupational demands, family needs and societal pressures. These influences may have additional direct and indirect relationships to cardiometabolic health separate from sleep. And these relationships are not resolved in experimental studies either if these social and environmental factors can predispose individuals to differential relationships between sleep health and cardiometabolic outcomes. This is important because interpretation of data showing links between sleep variables and cardiometabolic outcomes may reflect the contributions of other, unmeasured processes that are contributing to both (3).

Taken together, sleep is multidimensional yet much of the research to date has been unable to account for this complexity. This is important to recognise at the outset for several reasons. First, all published associations of sleep-related variables with cardiometabolic outcomes should be examined in light of the correlated elements of sleep health that may have been unmeasured or imperfectly measured in those studies (17). Given the interrelationship of elements of sleep health, an association with sleep duration, for example, may represent an association with sleep quality for which sleep duration may have been a proxy if that element of sleep quality was unmeasured.

Second, all interpretations of associations between sleep health and cardiometabolic health require the caveat that not only is sleep health a collection of interrelated processes that reflect both biological and environmental influences, but cardiometabolic health is also not a discrete variable. It represents the complex interrelationship of many physiologic processes that similarly reflect different dimensions of health and all correlate with each other. Given the complexity of cardiometabolic systems (e.g., processes such as inflammation, metabolism, endothelial function, sympathetic function, cardiac function), it is difficult to examine any specific outcome in isolation.

Third, single-timepoint measures or artificial acute situations may limit generalisability when discussing complex, dynamic processes such as sleep and cardiometabolic health. Both systems represent many environmental adaptations and include homeostatic and compensatory systems to adapt to changing environmental situations. Thus, sleep at one timepoint may predict health trajectories in the future, but that may depend on the trajectory of the sleep health variables themselves. Similarly, changes to cardiometabolic health outcomes may reflect changes to that particular system, but they may also reflect adapted responses to alterations in the environment. Thus, cardiometabolic outcomes may represent influence of the sleep variable itself, influence of related factors, compensation in the face of environmental pressures, etc. Current research paradigms are limited to relatively simplistic examinations of these processes due to limitations in budgets, time, technology and available scientific data regarding the complexities of the processes under study.

Although this chapter refers to specific associations, it should be noted that all these associations should be interpreted in light of these often-unmentioned caveats and qualifications. Future work is needed to disentangle these concepts. Still, the available evidence presents compelling reasons to believe that, despite these limitations to the data, important relationships exist.

Life's Essential 8

In 2010, the American Heart Association described a framework for conceptualising cardiovascular health, separately from cardiovascular disease. This framework, called 'Life's Simple 7' (LS7), attempted to describe the components of what it means to have cardiovascular health (18). The seven components included low blood pressure, healthy levels of blood lipids, low fasting glucose, healthy weight, healthy diet, sufficient physical activity and non-smoking. Over time, these seven components evolved into a general risk score for cardiovascular health and have been shown to predict cardiometabolic outcomes, including cardiovascular events and mortality. In the subsequent decade, emerging work from a number of fields motivated the reexamination of LS7. This culminated in 2022, when this framework was updated to 'Life's Essential 8' (LE8) (19). A few key changes made to the LS7 were (1) the components were conceptualised as being on a spectrum, rather than just 'good' or 'bad', (2) the components were conceptualised to be inextricably linked to contextual factors, especially social determinants of health, (3) the components were conceptualised to be influenced by psychological health and (4) the components were expanded to include sleep.

The inclusion of sleep health in the LE8 (19) officially elevates sleep as a component of cardiovascular health alongside more recognised factors such as high blood pressure and unhealthy diet and is an equal contributor to overall LE8 risk score. As described above, sleep is a multidimensional and complex set of processes. For the purposes of the LE8, though, scores are based on habitual sleep duration. This is not because habitual sleep duration is the best variable to approximate the role of sleep in cardiovascular health. Rather, it is because it is (a) the variable most studied and most consistently shown to be related to cardiometabolic outcomes of interest, (b) a variable that is amenable to both survey and experimental paradigms and (c) a variable that may serve as a proxy for other dimensions of sleep health if asked in isolation. Although an ideal representation of sleep health would be more nuanced, the inclusion of sleep duration in the LE8 should motivate additional work examining other dimensions of sleep health with the goal of clarifying the role of sleep in cardiometabolic health.

Recent studies show that adding sleep to the LE8 has improved the ability of the model to predict outcomes of interest. For example, sleep duration was shown to be related to the components of the LS7 on its own, but it also contributes to the prediction of overall health when it is added to the LS7 (20). Further, the individual contribution of sleep duration to overall health is similar in magnitude to other elements of the LS7. Another recent study showed that adding sleep to the LS7 to form the LE8 improved prediction of cardiovascular health trajectories (8).

Mechanisms linking sleep and cardiometabolic health

Typically, discussion of mechanisms of disease from a translational perspective involves genes, which transcribe proteins, which are expressed in cells, which exist in tissues, which play roles in some physiologic function; mechanisms, in these contexts, typically refer to the molecular pathways that influence the cellular processes that go awry in disease conditions. This chapter does not discuss these in great detail because (1) there is still very limited data on these pathways, and (2) a complete catalogue of all possible molecular mechanisms is outside the scope of this chapter. Rather, a big-picture view is taken. Figure 8.1 depicts such a big-picture view on the mechanisms linking sleep and cardiometabolic disease risk. In this figure, sleep health is represented as a single construct, although it encompasses a wide range of components, including sleep duration, sleep quality, sleep regularity, absence of sleep disorders and more. These dimensions of sleep health are hypothesised to work in concert with each other to impact health and functioning at the level of the brain as well as in the body.

Thus, the pathways leading from sleep health to cardiometabolic health are various, interactive and likely implicate a wide range of sleep-related functions in both the brain and the periphery. The following sections outline some of these in more detail.

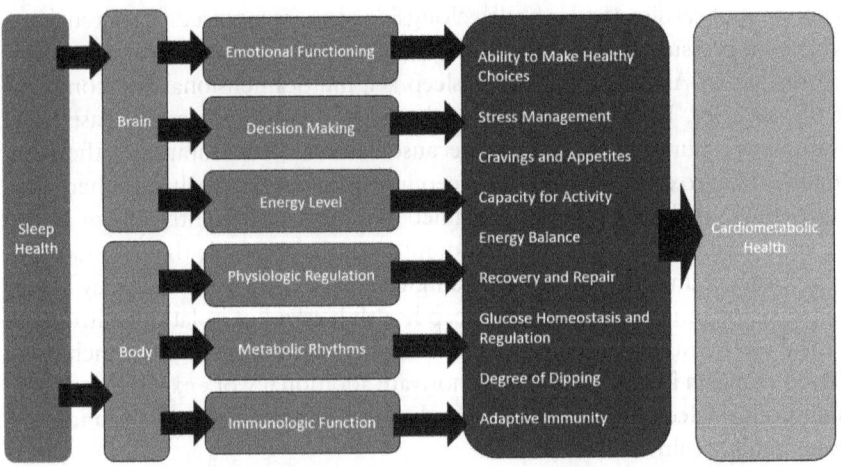

FIGURE 8.1   Schematic depicting various pathways linking sleep health and cardiometabolic health.

Mechanistic links between sleep health and obesity

The prevalence of obesity has increased to over 42% in the US (21). This is alarming, given many relationships between obesity and cardiometabolic disease (22). There are many pathways to obesity, and many pathways leading from obesity to cardiovascular and metabolic health conditions. These include both behavioural and physiologic components. Relationships between habitual sleep and obesity have been observed across many populations and have been summarised previously (23).

Sleep and food intake

Sleep health impacts obesity through several pathways, including perhaps the most proximal, food intake. Population-level studies show that obesogenic dietary patterns are more likely to occur in individuals with insufficient sleep, excessive sleep duration, poor sleep quality and sleep disorders (24). Specifically, studies repeatedly show that individuals in these groups are more likely to consume an unhealthy diet. However, this does not seem to be as simple as differences in caloric intake. Caloric intake does not robustly aggregate by sleep duration category in many studies (25, 26). This suggests that sleep may influence dietary quality more than quantity, or it may interact with other factors on caloric intake.

Several studies have shown that individuals who experience insufficient sleep and/or poor-quality sleep are more likely to engage in unhealthy eating patterns, including emotional eating (27), unrestricted eating, hedonic eating patterns (28) and later timing of meals (29). The latter is relevant in that several studies show that later timing of meals is associated with weight gain, even at comparable caloric intake (30, 31). This is further supported by time-restricted eating data that show that reducing the window of time that elapses between the start of eating and the last calories consumed each day is also associated with metabolic benefits in animals and humans (32).

Less sleep may simply allow for more time available to eat, leading to more calories consumed, especially at night. It may also alter decision-making processes around food. For example, many studies have documented the impact of sleep disruptions on decision-making (33). Applying these findings to food choices, it is plausible to suspect that the neurocognitive impacts of sleep disruption may interact with the metabolic ones. For example, existing data suggest that sleep-deprived individuals are more likely to report a desire to consume unhealthy foods, relative to the same individuals' well-rested state. In addition, the subjective discomfort associated with poor sleep quality or other sleep problems may lead to a desire to consume more pleasurable foods (34).

Some of these findings are supported by laboratory studies that show that when individuals experience a restricted sleep window in the laboratory, they consume about 350–550 additional calories, mostly at night after dinner (35). These studies uniformly show that acute sleep window reduction led to increases in consumption of food, especially at night. Acute weight changes from these conditions have also been documented. Other studies show that alterations in sleep can lead to changes in reward processing in the brain as it relates to food (36, 37).

Other studies from weight loss programs show relationships to sleep health. One study in a large workplace setting found that worse sleep quality was associated with an increased likelihood to engage in weight loss efforts but a reduced ability to maintain those efforts (38). Other studies show that individuals engaging in weight loss activities are more successful when they achieve sufficient sleep (39). Taken together, sleep may influence the quantity of food intake, quality of food intake and timing of food intake, which may all independently and in combination contribute to obesity.

Sleep health and exercise

The relationship between sleep health and exercise is complex. Several studies at the population level document this, suggesting that there is no clear relationship between energy expenditure and sleep duration and/or quality (40). Despite this lack of broad findings, there are several specific findings that link sleep with exercise.

Sleep is not a sedentary behaviour, despite its lack of movement. The idea of sleep as a non-sedentary part of the 24-hour activity cycle is a relatively recent conceptualisation. This understanding has changed how sleep is viewed within physical activity research, as a non-active but non-sedentary set of behaviours. Less sleep may afford individuals more opportunity to be active. While it may be the case that people are reducing sleep time to wake-up early enough to engage in exercise or other physically active behaviours, less healthy sleep may simultaneously lead to increased fatigue, lethargy and sleepiness (41). These would represent barriers to sufficient physical activity. Thus, there is a complex relationship whereby less sleep may both facilitate and interfere with physical activity. This is supported by laboratory studies that show that being kept awake is associated with increased caloric consumption during the portion of the day when the individual is typically asleep (35), mediating 24-hour differences in energy expenditure favouring sleep loss. In real-world situations, however, it is not clear whether this relationship is maintained and/or counteracted by increased tiredness at these same times.

Other studies show that fatigue (42) and daytime sleepiness (43) are inversely related to physical activity. Insufficient and excessive sleep duration, especially among later chronotypes, have been shown to be associated with increased daytime sedentary activity (44). Further, time-lagged research shows that a night of poor sleep may interfere with next-day exercise propensity (45). Clinically, people who experience poor sleep quality report less energy during the day, which supports these findings.

Mechanistic links between sleep health and hypertension

Sleep duration and hypertension

Many previous studies have documented an overlap between sleep duration and likelihood of hypertension (4, 46). Most of these studies were cross-sectional, indicating that individuals with shorter sleep duration (and, in some studies, longer sleep duration) were more likely to also have a history of hypertension. Additional studies have extended this work to document longitudinal associations, such that shorter sleepers are more likely to develop high blood pressure (47). Meta-analytic results indicate that habitual short sleep duration is associated with an approximately 20% increased likelihood of developing hypertension, compared to normal sleepers (48).

Mechanisms for this relationship have been explored but the causes of this relationship are likely multifactorial. Several candidates that have been previously studied in addition to obesity (described above) include impacts on 24-hour patterns of blood pressure regulation, impacts on sympathetic activation and impacts on stress.

Sleep health and 24-hour blood pressure

In addition to studies of cross-sectional relationships between sleep and hypertension incidence and prevalence, several studies have begun to explore relationships with 24-hour blood pressure. Unlike single-timepoint measures, 24-hour blood pressure measures do not have the same issues with readings that do not represent real-world levels. Often, blood pressure readings obtained in clinics are unreliable. The 24-hour recordings typically use devices that automatically obtain readings every 30 or 60 or 120 minutes across the 24-hour period. These values are then evaluated, with mean 24 hours, daytime and night-time values derived. In addition, this approach allows for the obtaining of blood pressure variability.

In work by Shulman and colleagues (49), two separate cohorts were examined. First, hypertensive patients enrolled in a treatment study were evaluated in the laboratory at baseline. Participants were not recruited based on their sleep duration at baseline, but based on a median split, shorter sleep was associated with elevated 24-hour blood pressure (including both daytime and night-time blood pressure), elevated 24-hour heart rate, and a greater rate of uncontrolled hypertension. Elevated 24-hour blood pressure was also seen in a cohort of sleep apnoea patients. In addition, both cohorts showed linear associations between sleep duration and 24-hour systolic blood pressure and 24-hour heart rate. Further, these relationships were maintained after adjusting for nocturnal dipping and in-office blood pressure values. Conversely, in an intervention study in the laboratory, sleep extension among habitual short sleepers of only about 30 minutes was associated with a decrease in mean blood pressure. These data suggest that sleep duration directly or indirectly impacts the 24-hour regulation of blood pressure.

Sleep health and sympathetic activation

The role of sympathetic activation in blood pressure regulation is complex and involves organs in addition to the heart, including the kidneys, carotid body and several components of the endocrine system. Direct effects from the brain to the kidney implicate the renin-angiotensin system, whereas indirect effects may implicate hemodynamic changes mediated by sympathetic activation. Although the molecular mechanisms involved in this pathway have been well-described, the influence of sleep health on these has not.

Previous studies have shown that, in general, sympathetic nerve activity is lowest in slow wave sleep and highest in REM sleep, and this is accompanied by concomitant changes to beat-to-beat blood pressure (50). Further, K-complexes resulting from environmental stimuli (but not sleep spindles) produced associated bursts of sympathetic activity and blood pressure. Animal studies suggest that these changes may be mediated by activities in pontine nuclei (51). Although insomnia has been consistently linked with hyperarousal (52), and hyperarousal has been previously shown to have mechanistic links to sympathetic activation, detailed mapping of the pathways linking insomnia to sympathetic activation has not yet been completely successful. Some studies show associations with hypothalamic–pituitary–adrenal axis (53), growth hormone, cortisol (54) and other systems linked with sympathetic activation.

Several studies have also shown that insufficient sleep and/or insomnia can be associated with alterations in catecholamine levels (55). For example, sleep deprivation has been associated with elevated plasma norepinephrine. Additional work suggests that this is exacerbated by concurrent circadian disruption. Several studies have shown that insomnia contributes to sympathetic activation via norepinephrine as well (56).

Mechanistic links between sleep health and dyslipidaemia

There is somewhat of a paradoxical relationship between sleep health and dyslipidaemia. Unlike obesity, hypertension, and diabetes, where overlaps with various aspects of sleep health are relatively robust (as evidenced in meta-analyses), relationships to cholesterol and dyslipidaemia are less so. Several studies document relationships between sleep health and lipid health (57–60). However, prospective studies do not seem to support a causal (or at least temporal) association. Thus, it may be easy to conclude that sleep health has less of an impact on lipids than other domains of cardiometabolic health.

However, the relationship seems more complex. First, there are a breadth of studies that have documented associations between sleep health and aspects of atherosclerosis risk (61–63), which is largely driven by lipid health. Further, there are several studies that show that cholesterol synthesis in the liver is highly sleep dependent (64, 65). The genes that transcribe the proteins involved in the cholesterol synthesis pathway are nearly universally upregulated during sleep. In other words, cholesterol synthesis depends on sleep.

This may explain some of the complex findings. The cross-sectional analyses may find relationships due to overlap with other cardiometabolic conditions; since those with higher cholesterol typically also exhibit other cardiometabolic risks which might be more reliably linked with sleep health, that may explain some of the overlap. However, if cholesterol synthesis requires sleep, less sleep may actually inhibit cholesterol synthesis. In this case, there would be two influences: concurrent risk factors elevating risk of dyslipidaemia and reduced sleep reducing lipid production at the same time. This may also explain why associations between sleep duration and some elements of atherosclerosis risk are more prominent in excessive sleep rather than insufficient sleep. More work is needed to better understand this complex relationship.

Mechanistic links between sleep health and diabetes

Sleep and diabetes physiology

During sleep, many physiologic processes are restored, including modulation of metabolic and endocrine systems. Under normal conditions, metabolic rate decreases and growth hormone (GH) increases during non-REM sleep. GH is well known to induce insulin resistance (66), and the release of GH during non-REM sleep likely plays a reparative role. However, circadian misalignment (67) and poor sleep have been shown to cause significant elevated glucose and insulin levels, and insulin resistance (68, 69). Additionally, longitudinal and experimental studies have shown that decreasing sleep duration is associated with dysregulated glucose homeostasis. Rafalson et al. (70) re-examined participants who were free of cardiometabolic disease at baseline and found that short sleepers were three times as likely to have impaired fasting glucose compared to regular sleepers. These findings have been supported by carefully controlled laboratory studies where researchers examined the relationship between sleep restriction and glucose regulation and found that after only six nights of sleep restriction to four hours per night, participants were 40% less effective at regulating blood glucose (71). Through both decreased insulin release and increased cellular insulin resistance, short sleep duration has been shown to impact glucose regulation.

Sleep and metabolic hormones

Sleep also impacts metabolic hormones that regulate appetite, which may play a role in development of obesity and subsequent Type-2 Diabetes Mellitus (T2DM). Studies have shown that sleep loss may be involved in the reduction of leptin and increase of ghrelin. Leptin, the satiety-signalling hormone, negatively modulates the food reward system by chemically indicating satiety (72), and conversely, ghrelin positively modulates the system (73) by chemically indicating feelings of hunger. Taken together, sleep loss appears to negatively impact homeostatic eating through decreased leptin and increased ghrelin concentrations, ultimately decreasing feelings of satiety and increasing feelings of hunger. In addition to greater feelings of hunger following short sleep, short sleep has also been shown to be associated with increased appetite (34). Furthermore, insufficient sleep is associated with an increase in low-grade inflammation and proinflammatory cytokines, which may increase insulin (74, 75) and leptin resistance (76). These disruptions to the food reward system result in hedonic eating, increased caloric consumption, and may lead to obesity, a known risk factor for T2DM.

Diabetes and insomnia

It has been estimated that nearly 40% of individuals with T2DM have insomnia or insomnia symptoms (77). The relationship between insomnia and T2DM is likely bidirectional. LeBlanc et al. (78) found that persons with insomnia were 28% more likely to develop T2DM. This is likely explained by the metabolic dysregulating effects of insufficient sleep. Additionally, persons with T2DM may be less likely to fall and stay asleep for several reasons. First, hyperglycaemia (high blood sugar) can cause increased thirst and urination, resulting in disrupted sleep and more frequent trips to the bathroom throughout the night. Additionally, uncontrolled blood glucose can result in feelings of hunger, which in turn may result in shorter sleep duration and later bedtimes. Overall, insomnia and T2DM commonly coexist and if left untreated may continue to exacerbate one another.

Obstructive sleep apnoea (OSA) and diabetes

Obstructive sleep apnoea (OSA), a chronic treatable sleep disorder, is a common comorbidity in persons with T2DM. OSA can cause intermittent hypoxemia (low blood oxygen levels) and hypercapnia (elevated arterial carbon dioxide tension), increased oxidative stress, inflammation, and fragmented sleep, all of which are important mediators of metabolic risk (79). Several cross-sectional studies have shown that OSA severity is independently associated with insulin resistance in the absence of T2DM (80–82), indicating that OSA may predispose T2DM. More research is needed to determine whether poorly controlled T2DM can worsen OSA by adversely impacting central control of respiration; however, studies have demonstrated a high prevalence of OSA in non-obese persons with type 1 diabetes mellitus (83–85), suggesting that reverse causality may exist. Furthermore, like T2DM, obesity appears to be a risk factor for OSA (86–88), highlighting an important modifiable risk factor for both OSA and T2DM.

Clinical considerations

Understanding the connection between sleep and cardiovascular disease can help sleep medicine providers in several ways:

1 Early identification of individuals at risk: Sleep medicine providers can identify individuals at risk of developing cardiovascular disease based on their sleep patterns and provide early interventions to prevent the onset of the disease.
2 Tailored treatment: By understanding the connection between sleep and cardiovascular disease, sleep medicine providers can develop tailored treatment plans to improve their patients' sleep patterns, thereby reducing their risk of developing cardiovascular disease.
3 Increased patient awareness: Sleep medicine providers can educate their patients about the connection between sleep and cardiovascular disease, which can motivate patients to adopt healthy sleep habits and lifestyle changes that can reduce their risk of developing cardiovascular disease. Cognitive Behavioral Therapy for Insomnia (CBT-I) could potentially be modified to also include psychoeducation about cardiovascular diseases. They also can introduce the idea that insufficient sleep can negatively impact one's ability to make good decisions about diet and physical activity, which also negatively impacts cardiovascular health (89).
4 Improved patient outcomes: By addressing sleep issues in patients with cardiovascular disease, sleep medicine providers can improve patient outcomes by reducing the risk of further complications and improving overall health and quality of life. For example, short sleep has been associated with decreased leptin and increased ghrelin levels, which negatively impacts subsequent satiety. By addressing insufficient sleep directly, the provider may prevent the ensuing weight gain, suboptimal eating patterns, and energy imbalance that is a direct result of short sleep and may further harm cardiovascular health (90, 91).

Understanding the connection between sleep and cardiovascular disease can be helpful for physicians in several ways:

1 Identifying at-risk patients: Physicians can use knowledge of the link between sleep and cardiovascular disease to identify patients who may be at higher risk for developing heart disease. Since patients with OSA may be at

higher risk for hypertension and other cardiovascular problems, physicians can monitor these patients more closely and make targeted referrals for coordinated care with sleep medicine.

2 Recommending lifestyle changes: Physicians treating patients with cardiovascular disease may address LE8 with regard to sleep by including sleep as part of their assessment and treatment program. It is key to highlight to patients that sufficient sleep is an important part of preserving cardiovascular health and/or improving suboptimal cardiovascular health (19). Physicians may recommend that patients prioritise improved sleep health practices, such as going to bed and waking up at the same time every day, avoiding caffeine and alcohol before bedtime, and creating a relaxing sleep environment. By addressing sleep directly, physicians can improve their ability to successfully treat the cardiovascular condition (e.g., hypertension). They also can introduce the idea that insufficient sleep can negatively impact one's ability to make good decisions about diet and physical activity, which also negatively impacts cardiovascular health (89).

3 Prescribing treatments: Finally, knowing the link between sleep and cardiovascular disease can help physicians determine the best course of treatment for patients who are already experiencing heart problems. For example, doctors may refer to a sleep medicine physician to determine if patients with treatment-resistant hypertension may meet criteria for OSA. If OSA is present, the sleep medicine physician can prescribe treatments like positive airway pressure (PAP) machines or oral appliances to help patients with sleep apnoea breathe more easily during the night. This referral can lead to a diagnostic process and subsequent treatment approach that can reduce a patient's risk of hypertension and other cardiovascular problems, and potentially improve existing suboptimal cardiovascular health (92).

Future directions

Although sleep health has been recognised as a core component of cardiometabolic health, much work is still needed to understand, clarify, and expand these findings. First, better mechanistic studies linking sleep and outcomes are needed. These will help delineate the specific molecular and behavioural pathways, rather than simply document associations. This work will not only add clarity to the discussion, but it may uncover novel intervention targets for ameliorating the risks associated with insufficient sleep.

Second, future work should better understand the role of sleep and cardiometabolic health in context. Sleep is a highly contextualised phenomenon and is determined by a broad range of individual, social, and societal factors. Better understanding the role of these in the relationship between sleep and cardiometabolic health will be useful for better understanding the role of sleep health as part of the overall picture of health.

Third, future work needs to better clarify the causal pathways linking various aspects of sleep health and various aspects of cardiometabolic health. Since many elements of sleep correlate with each other and many cardiometabolic risks cause each other, a better understanding of which elements are driving which other elements will improve the ability of interventions to ameliorate risk. For example, if the primary reason that sleep is linked with cardiometabolic risk is through metabolic pathways, and the main culprit is eating behaviours, interventions targeting eating behaviours may be the most beneficial.

Conclusions

Sleep is a fundamental set of physiologic phenomena that exist in a contextualised framework. As such, sleep health (and its various components) plays an important role in cardiometabolic health. Insufficient sleep duration, poor sleep quality and other elements of disturbed sleep and sleep disorders are associated with changes in both the brain and the body. These changes result in alterations in emotional functioning, decision-making, energy level, physiologic regulation, metabolic rhythms and immunologic function. These alterations then result in consequences in outcomes such as ability to make healthy choices, management of stress, experience of cravings and appetites, capacity for physical activity, ability to manage energy balance, capacity for recovery and repair, homeostasis and regulation of glucose and insulin, degree of nocturnal dipping of blood pressure and heart rate and ability of the immune system to successfully adapt to the environment. These then interact to contribute to the molecular and mechanistic pathways that directly impact weight gain and obesity, blood pressure and hypertension, cholesterol and dyslipidaemia, insulin resistance and diabetes, and other elements of cardiometabolic health.

References

1 Worley SL. The extraordinary importance of sleep: the detrimental effects of inadequate sleep on health and public safety drive an explosion of sleep research. P T. 2018;43(12):758–63.
2 American Psychiatric Association. Diagnostic and Statistical Manual of Mental Disorders: 5th Edition: DSM-5. American Psychiatric Association; 2003. Available from: https://books.google.com/books?id=obZknQEACAAJ
3 Grandner MA, Fernandez FX. The translational neuroscience of sleep: a contextual framework. Science. 2021;374(6567):568–73.
4 Grandner M, Mullington JM, Hashmi SD, Redeker NS, Watson NF, Morgenthaler TI. Sleep duration and hypertension: analysis of > 700,000 adults by age and sex. J Clin Sleep Med. 2018;14(6):1031–9.
5 Reutrakul S, Van Cauter E. Sleep influences on obesity, insulin resistance, and risk of type 2 diabetes. Metabolism. 2018;84:56–66.

6 Irwin MR, Olmstead R, Carroll JE. Sleep disturbance, sleep duration, and inflammation: a systematic review and meta-analysis of cohort studies and experimental sleep deprivation. Biol Psychiatry. 2016;80(1):40–52.

7 Irwin MR, Vitiello MV. Implications of sleep disturbance and inflammation for Alzheimer's disease dementia. Lancet Neurol. 2019;18(3):296–306.

8 Makarem N, Castro-Diehl C, St-Onge MP, Redline S, Shea S, Lloyd-Jones D, et al. Redefining cardiovascular health to include sleep: prospective associations with cardiovascular disease in the MESA sleep study. J Am Heart Assoc. 2022;11(21):e025252.

9 Buysse DJ. Sleep health: can we define it? Does it matter? Sleep. 2014;37(1):9–17.

10 Grandner MA. Sleep, health, and society. Sleep Med Clin. 2022;17(2).

11 Grandner MA, Patel NP, Gehrman PR, Perlis ML, Pack AI. Problems associated with short sleep: bridging the gap between laboratory and epidemiological studies. Sleep Med Rev. 2010;14(4).

12 Watson NF, Badr MS, Belenky G, Bliwise DL, Buxton OM, Buysse D, et al. Joint consensus statement of the American academy of sleep medicine and sleep research society on the recommended amount of sleep for a healthy adult: methodology and discussion. Sleep. 2015;38(8):1161–83.

13 Watson NF, Badr MS, Belenky G, Bliwise DL, Buxton OM, Buysse D, et al. Recommended amount of sleep for a healthy adult: a joint consensus statement of the American academy of sleep medicine and sleep research society. Sleep. 2015;38(6):843–4.

14 Ferrie JE, Kivimaki M, Akbaraly TN, Singh-Manoux A, Miller MA, Gimeno D, et al. Associations between change in sleep duration and inflammation: findings on C-reactive protein and interleukin 6 in the Whitehall II Study. Am J Epidemiol. 2013;178(6):956–61.

15 Perlis ML, Pigeon WR, Grandner MA, Bishop TM, Riemann D, Ellis JG, et al. Why treat insomnia? J Prim Care Community Health. 2021;12.

16 Baltzis D, Bakker JP, Patel SR, Veves A. Obstructive sleep apnea and vascular diseases. Compr Physiol. 2016;6(3):1519–28.

17 Patterson F, Brewer B, Blair R, Grandner MA, Hoopes E, Ma G, et al. An exploration of clinical, behavioral, and community factors associated with sleep duration and efficiency among middle-aged Black/African American smokers. Sleep Health. 2021;7(3).

18 Lloyd-Jones DM, Hong Y, Labarthe D, Mozaffarian D, Appel LJ, Van Horn L, et al. Defining and setting national goals for cardiovascular health promotion and disease reduction: the American heart association's strategic impact goal through 2020 and beyond. Circulation. 2010;121(4):586–613.

19 Lloyd-Jones DM, Allen NB, Anderson CAM, Black T, Brewer LC, Foraker RE, et al. Life's essential 8: updating and enhancing the American heart association's construct of cardiovascular health: a presidential advisory from the American heart association. Circulation. 2022;146(5).

20 Grandner MA, Patterson F, Malone SK, Hanlon A, Haynes P, Petrov ME, et al. Should habitual sleep duration be added to "life's simple 7?" Circulation. 2016;133(Suppl 1):AMP95.

21 Fryar CD, Carroll MD, Afful J. Prevalence of overweight, obesity, and severe obesity among adults aged 20 and over: United States, 1960–1962 through 2017–2018. NCHS Health E-Stats, Centers for Disease Control and Prevention. 2020.

22 Drozdz D, Alvarez-Pitti J, Wójcik M, Borghi C, Gabbianelli R, Mazur A, et al. Obesity and cardiometabolic risk factors: from childhood to adulthood. Nutrients. 2021;13(11).

23 Beccuti G, Pannain S. Sleep and obesity. Curr Opin Clin Nutr Metab Care. 2011;14(4):402–12.

24 Chaput JP, Bouchard C, Tremblay A. Change in sleep duration and visceral fat accumulation over 6 years in adults. Obesity (Silver Spring). 2014. Available from: www.ncbi.nlm.nih.gov/pubmed/24420871

25 Grandner MA, Jackson N, Gerstner JR, Knutson KL. Dietary nutrients associated with short and long sleep duration: data from a nationally representative sample. Appetite. 2013;64:71–80.

26 Grandner MA, Jackson N, Gerstner JR, Knutson KL. Sleep symptoms associated with intake of specific dietary nutrients. J Sleep Res. 2014;23(1).

27 Palmer CA, Alfano CA. Sleep and emotion regulation: an organizing, integrative review. Sleep Med Rev. 2017;31:6–16.

28 Dashti HS, Scheer FA, Jacques PF, Lamon-Fava S, Ordovas JM. Short sleep duration and dietary intake: epidemiologic evidence, mechanisms, and health implications. Adv Nutr. 2015;6(6):648–59.

29 Phoi YY, Rogers M, Bonham MP, Dorrian J, Coates AM. A scoping review of chronotype and temporal patterns of eating of adults: tools used, findings, and future directions. Nutr Res Rev. 2022;35(1):112–35.

30 Davis R, Rogers M, Coates AM, Leung GKW, Bonham MP. The impact of meal timing on risk of weight gain and development of obesity: a review of the current evidence and opportunities for dietary intervention. Curr Diab Rep. 2022;22(4):147–55.

31 Lopez-Minguez J, Gómez-Abellán P, Garaulet M. Timing of breakfast, lunch, and dinner. Effects on obesity and metabolic risk. Nutrients. 2019;11(11).

32 Boyd P, O'Connor SG, Heckman-Stoddard BM, Sauter ER. Time-restricted feeding studies and possible human benefit. JNCI Cancer Spectr. 2022;6(3).

33 Harrison Y, Horne JA. Should we be taking more sleep? Sleep. 1995;18(10):901–7.

34 Spiegel K, Tasali E, Penev P, Van Cauter E. Brief communication: sleep curtailment in healthy young men is associated with decreased leptin levels, elevated ghrelin levels, and increased hunger and appetite. Ann Intern Med. 2004;141(11):846–50.

35 Spaeth AM, Dinges DF, Goel N. Effects of experimental sleep restriction on weight gain, caloric intake, and meal timing in healthy adults. Sleep. 2013;36(7):981–90.

36 Duraccio KM, Krietsch KN, Zhang N, Whitacre C, Howarth T, Pfeiffer M, et al. The impact of short sleep on food reward processes in adolescents. J Sleep Res. 2021;30(2):e13054.

37 Rihm JS, Menz MM, Schultz H, Bruder L, Schilbach L, Schmid SM, et al. Sleep deprivation selectively upregulates an amygdala-hypothalamic circuit involved in food reward. J Neurosci. 2019;39(5):888–99.

38 Hui SKA, Grandner MA. Associations between poor sleep quality and stages of change of multiple health behaviors among participants of employee wellness program. Prev Med Rep. 2015;2.

39 Papatriantafyllou E, Efthymiou D, Zoumbaneas E, Popescu CA, Vassilopoulou E. Sleep deprivation: effects on weight loss and weight loss maintenance. Nutrients. 2022;14(8).

40 St-Onge MP. The role of sleep duration in the regulation of energy balance: effects on energy intakes and expenditure. J Clin Sleep Med. 2013;9(1):73–80.

41 Nelson V, Dubov A, Morton K, Fraenkel L. Using nominal group technique among resident physicians to identify key attributes of a burnout prevention program. PLoS ONE. 2022;17(3):e0264921.

42 Wender CLA, Manninen M, O'Connor PJ. The effect of chronic exercise on energy and fatigue states: a systematic review and meta-analysis of randomized trials. Front Psychol. 2022;13:907637.

43 McClain JJ, Lewin DS, Laposky AD, Kahle L, Berrigan D. Associations between physical activity, sedentary time, sleep duration and daytime sleepiness in US adults. Prev Med (Baltim). 2014;66:68–73.

44 Patterson F, Malone SK, Lozano A, Grandner MA, Hanlon AL. Smoking, screen-based sedentary behavior, and diet associated with habitual sleep duration and chronotype: data from the UK biobank. Ann Behav Med. 2016;50(5).

45 Kline CE. The bidirectional relationship between exercise and sleep: implications for exercise adherence and sleep improvement. Am J Lifestyle Med. 2014;8(6):375–9.

46 Gangwisch JE, Heymsfield SB, Boden-Albala B, Buijs RM, Kreier F, Pickering TG, et al. Short sleep duration as a risk factor for hypertension: analyses of the first National Health and Nutrition Examination Survey. Hypertension. 2006;47(5):833–9.

47 Calhoun DA, Harding SM. Sleep and hypertension. Chest. 2010;138(2):434–43.

48 Meng L, Zheng Y, Hui R. The relationship of sleep duration and insomnia to risk of hypertension incidence: a meta-analysis of prospective cohort studies. Hypertens Res. 2013;36(11):985–95.

49 Shulman R, Cohen DL, Grandner MA, Gislason T, Pack AI, Kuna ST, et al. Sleep duration and 24-hour ambulatory blood pressure in adults not on antihypertensive medications. J Clin Hypertens. 2018;20(12).

50 Somers VK, Dyken ME, Mark AL, Abboud FM. Sympathetic-nerve activity during sleep in normal subjects. N Engl J Med. 1993;328(5):303–7.

51 Fink AM, Burke LA, Sharma K. Lesioning of the pedunculopontine nucleus reduces rapid eye movement sleep, but does not alter cardiorespiratory activities during sleep, under hypoxic conditions in rats. Respir Physiol Neurobiol. 2021;288:103653.

52 Kalmbach DA, Cuamatzi-Castelan AS, Tonnu CV, Tran KM, Anderson JR, Roth T, et al. Hyperarousal and sleep reactivity in insomnia: current insights. Nat Sci Sleep. 2018;10:193–201.

53 Dressle RJ, Feige B, Spiegelhalder K, Schmucker C, Benz F, Mey NC, et al. HPA axis activity in patients with chronic insomnia: a systematic review and meta-analysis of case-control studies. Sleep Med Rev. 2022;62:101588.

54 Van Cauter E, Spiegel K, Tasali E, Leproult R. Metabolic consequences of sleep and sleep loss. Sleep Med. 2008;9(Suppl 1):S23–8.

55 Greenlund IM, Carter JR. Sympathetic neural responses to sleep disorders and insufficiencies. Am J Physiol Heart Circ Physiol. 2022;322(3):H337–49.

56 Mitchell HA, Weinshenker D. Good night and good luck: norepinephrine in sleep pharmacology. Biochem Pharmacol. 2010;79(6):801–9.

57 Kong AP, Wing YK, Choi KC, Li AM, Ko GT, Ma RC, et al. Associations of sleep duration with obesity and serum lipid profile in children and adolescents. Sleep Med. 2011;12(7):659–65.

58 Smiley A, King D, Harezlak J, Dinh P, Bidulescu A. The association between sleep duration and lipid profiles: the NHANES 2013–2014. J Diabetes Metab Disord. 2019;18(2):315–22.

59 Zhan Y, Chen R, Yu J. Sleep duration and abnormal serum lipids: the China Health and Nutrition Survey. Sleep Med. 2014;15(7):833–9.

60 Altman NG, Izci-Balserak B, Schopfer E, Jackson N, Rattanaumpawan P, Gehrman PR, et al. Sleep duration versus sleep insufficiency as predictors of cardiometabolic health outcomes. Sleep Med. 2012;13(10).

61 Pan XL, Nie L, Zhao SY, Zhang XB, Zhang S, Su ZF. The association between insomnia and atherosclerosis: a brief report. Nat Sci Sleep. 2022;14:443–8.

62 Full KM, Pusalavidyasagar S, Palta P, Sullivan KJ, Shin JI, Gottesman RF, et al. Associations of late-life sleep medication use with incident dementia in the atherosclerosis risk in communities study. J Gerontol A Biol Sci Med Sci. 2023;78(3):438–46.

63 Kadoya M, Koyama H. Sleep, autonomic nervous function and atherosclerosis. Int J Mol Sci. 2019;20(4).

64 Aho V, Ollila HM, Kronholm E, Bondia-Pons I, Soininen P, Kangas AJ, et al. Prolonged sleep restriction induces changes in pathways involved in cholesterol metabolism and inflammatory responses. Sci Rep. 2016;6:24828.

65 Xing C, Huang X, Zhang Y, Zhang C, Wang W, Wu L, et al. Sleep disturbance induces increased cholesterol level by NR1D1 mediated CYP7A1 inhibition. Front Genet. 2020;11:610496.

66 Møller N, Jørgensen JOL. Effects of growth hormone on glucose, lipid, and protein metabolism in human subjects. Endocr Rev. 2009;30(2):152–77.

67 Bass J, Takahashi JS. Circadian integration of metabolism and energetics. Science. 2010;330(6009):1349–54.

68 Knutson KL, Van Cauter E, Zee P, Liu K, Lauderdale DS. Cross-sectional associations between measures of sleep and markers of glucose metabolism among subjects with and without diabetes: the Coronary Artery Risk Development in Young Adults (CARDIA) Sleep Study. Diabetes Care. 2011;34(5):1171–6. Available from: www.ncbi.nlm.nih.gov/pubmed/21411507

69 Jennings JR, Muldoon MF, Hall M, Buysse DJ, Manuck SB. Self-reported sleep quality is associated with the metabolic syndrome. Sleep. 2007;30(2):219–23.

70 Rafalson L, Donahue RP, Stranges S, Lamonte MJ, Dmochowski J, Dorn J, et al. Short sleep duration is associated with the development of impaired fasting glucose: the Western New York Health Study. Ann Epidemiol. 2010;20(12):883–9.

71 Spiegel K, Leproult R, Van Cauter E. Impact of sleep debt on metabolic and endocrine function. Lancet. 1999;354(9188):1435–9.

72 Davis C. A good night's sleep. Nurs Stand. 2010;24(21):20–1.

73 Lutter M, Nestler EJ. Homeostatic and hedonic signals interact in the regulation of food intake. J Nutr. 2009;139(3):629–32.

74 Vgontzas AN, Zoumakis E, Bixler EO, Lin HM, Follett H, Kales A, et al. Adverse effects of modest sleep restriction on sleepiness, performance, and inflammatory cytokines. J Clin Endocrinol Metab. 2004;89(5):2119–26.

75 van Leeuwen WM, Hublin C, Sallinen M, Harma M, Hirvonen A, Porkka-Heiskanen T. Prolonged sleep restriction affects glucose metabolism in healthy young men. Int J Endocrinol. 2010;2010:108641.

76 Chen K, Li F, Li J, Cai H, Strom S, Bisello A, et al. Induction of leptin resistance through direct interaction of C-reactive protein with leptin. Nat Med. 2006;12(4):425–32.

77 Koopman ADM, Beulens JW, Dijkstra T, Pouwer F, Bremmer MA, van Straten A, et al. Prevalence of insomnia (symptoms) in T2D and association with metabolic parameters and glycemic control: meta-analysis. J Clin Endocrinol Metab. 2020;105(3):614–43.

78 LeBlanc ES, Smith NX, Nichols GA, Allison MJ, Clarke GN. Insomnia is associated with an increased risk of type 2 diabetes in the clinical setting. BMJ Open Diabetes Res Care. 2018;6(1):e000604.

79 Heinzer R, Vat S, Marques-Vidal P, Marti-Soler H, Andries D, Tobback N, et al. Prevalence of sleep-disordered breathing in the general population: the HypnoLaus study. Lancet Respir Med. 2015;3(4):310–18.

80 Punjabi NM, Sorkin JD, Katzel LI, Goldberg AP, Schwartz AR, Smith PL. Sleep-disordered breathing and insulin resistance in middle-aged and overweight men. Am J Respir Crit Care Med. 2002;165(5):677–82.

81 Punjabi NM. Do sleep disorders and associated treatments impact glucose metabolism? Drugs. 2009;69(Suppl 2):13–27.

82 Ip MS, Lam B, Lauder IJ, Tsang KW, Chung KF, Mok YW, et al. A community study of sleep-disordered breathing in middle-aged Chinese men in Hong Kong. Chest. 2001;119(1):62–9.

83 Banghoej AM, Nerild HH, Kristensen PL, Pedersen-Bjergaard U, Fleischer J, Jensen AEK, et al. Obstructive sleep apnoea is frequent in patients with type 1 diabetes. J Diabetes Complicat. 2017;31(1):156–61.

84 Borel AL, Benhamou PY, Baguet JP, Halimi S, Levy P, Mallion JM, et al. High prevalence of obstructive sleep apnoea syndrome in a Type 1 diabetic adult population: a pilot study. Diabet Med. 2010;27(11):1328–9.

85 Reutrakul S, Thakkinstian A, Anothaisintawee T, Chontong S, Borel AL, Perfect MM, et al. Sleep characteristics in type 1 diabetes and associations with glycemic control: systematic review and meta-analysis. Sleep Med. 2016;23:26–45.

86 Durán J, Esnaola S, Rubio R, Iztueta A. Obstructive sleep apnea-hypopnea and related clinical features in a population-based sample of subjects aged 30 to 70 yr. Am J Respir Crit Care Med. 2001;163(3 Pt 1):685–9.

87 Young T, Peppard PE, Gottlieb DJ. Epidemiology of obstructive sleep apnea: a population health perspective. Am J Respir Crit Care Med. 2002;165(9):1217–39.

88 Peppard PE, Young T, Barnet JH, Palta M, Hagen EW, Hla KM. Increased prevalence of sleep-disordered breathing in adults. Am J Epidemiol. 2013;177(9):1006–14.

89 Chattu VK, Chattu SK, Burman D, Spence DW, Pandi-Perumal SR. The inter-linked rising epidemic of insufficient sleep and diabetes mellitus. Healthcare (Basel). 2019;7(1).

90 St-Onge MP, Roberts AL, Chen J, Kelleman M, O'Keeffe M, RoyChoudhury A, et al. Short sleep duration increases energy intakes but does not change energy expenditure in normal-weight individuals. Am J Clin Nutr. 2011;94(2):410–16.

91 Grandner MA, Alfonso-Miller P, Fernandez-Mendoza J, Shetty S, Shenoy S, Combs D. Sleep: important considerations for the prevention of cardiovascular disease. Curr Opin Cardiol. 2016;31(5).

92 Drager LF, Lee CH. Treatment of obstructive sleep apnoea as primary or second-ary prevention of cardiovascular disease: where do we stand now? Curr Opin Pulm Med. 2018;24(6):537–42.

9

PAIN AND SLEEP

Underlying mechanisms of the sleep-pain relationship

Thomas Bilterys, Jo Nijs and Nicole Tang

Acknowledgments: This project has received funding from the European Union's Horizon 2020 research and innovation programme under the Marie Skłodowska-Curie grant agreement No 945380

Background

Pain, as defined by the International Association for the Study of Pain (IASP), is an unpleasant sensory and emotional experience associated with actual or potential tissue damage or described in terms of such damage (1). The primary function of pain in the human body is to warn us in case of (potential) damage to our body. In many circumstances, pain makes us take actions to avoid (potential) damage and ensures that our body can recover in time. However, sometimes pain persists for a longer period past the 'acute danger period', is not in proportion with the actual injury or the potential damage (anymore), and loses its 'warning' function (2). This is the case in many people with chronic pain, where pain frequently reoccurs over time or persists for a long time beyond the normal tissue healing time (3). They often have an increased responsiveness in central pain pathways, which results in a stronger pain response on noxious stimuli. Furthermore, some chronic pain patients continue to feel pain even though the initial injury is completely healed without any sign of noxious stimuli or (potential) damage (4, 5). In general, chronic pain has a large impact on patients' personal life due to interference with daily activities and social life, a negative influence on mental well-being and high medical costs (6, 7). It is also one of the most common reasons why people seek medical care (8). Several chronic pain conditions are the leading causes of disability, with chronic back pain being the number one disabling condition

DOI: 10.4324/9781003296966-9

worldwide, considering the prevalence and disability weight (9). Furthermore, chronic pain also has a detrimental effect on patients, their social and family environment, as well as on health care services and general economic well-being (6, 7). With reported prevalence rates of 11–40%, an estimated point prevalence of 20.4% and the significant impact on both the individual and the society, chronic pain can be considered a major public health problem (10).

Several decades ago, pain research mainly focused on changes on a biological level such as changes in sensory modalities, neurological transmission or other neurophysiological aspects (11). During this time, anxiety, depression, stress and sleep problems in people with chronic pain were often considered consequences or symptoms of chronic pain. However, gradually, the biopsychosocial model received more attention and people started to realise that psychosocial factors play a more important role in the development and persistence of pain. In the biopsychosocial model, pain is viewed as a dynamic interaction among biological, psychological and social factors that reciprocally influence each other (11, 12). This also implies that the process of experiencing pain is not just a result of a transfer of sensory input to the brain but rather a complex phenomenon where signals can be modified through inhibitory and excitatory systems and mechanisms.

The relationship between sleep and pain

One aspect that gained increasing attention in the pain literature and which appears to play an important role in chronic pain is sleep. Among people with chronic pain, sleep problems are very prevalent, with 53%–90% reporting difficulties with initiating sleep and frequent awakenings or having unrefreshing sleep (13, 14). Although the association between poor quality and chronic pain seems obvious, its directionality has been debated for over a decade. Initially, it was thought that sleep problems were a result of chronic pain. However, the available literature suggests that sleep disturbances are more than symptoms of chronic pain but rather a driver of a bidirectional sleep-pain relationship. Furthermore, it seems that sleep is a stronger predictor of pain than vice versa (15–18). Evidence regarding the association between sleep and pain even suggests that sleep is a predisposing factor for the onset and worsening of chronic pain conditions (19). Several sleep (deprivation) studies indicate that poor sleep can lead to lower pain thresholds and higher pain sensitivity the following day (20–22). Longitudinal studies showed that sleep impairments can predict the onset and exacerbation of chronic pain and lower quality of life (15, 23, 24). A microlongitudinal study indicated that in people with chronic pain and concomitant insomnia, presleep pain was not a reliable predictor of subsequent sleep, but sleep was better predicted by presleep cognitive arousal (25). Moreover, sleep quality was a consistent predictor of pain the next day (25). Especially, during the first half of the day, a pain-relieving effect of sleep could be observed (25). Furthermore, chronic pain patients appear to spontaneously engage in more physical activity following a better night of sleep (26).

Insomnia also seems to be associated with high pain intensity, anxiety, depression, pain catastrophising, comorbidities, healthcare use, perceived impact of pain on daily functioning and life satisfaction in people with chronic spinal pain (27, 28). The bidirectional relation could lead to a vicious cycle as pain itself can negatively influence and disrupt sleep and sleep disturbances can increase pain sensitivity. Consequently, there is a need to better understand both sleep and pain processes, their interaction, and their mechanisms.

Mechanisms of the sleep-pain relationship

Despite the fact that the link between sleep and pain has been broadly discussed in the literature, the underlying mechanisms of the sleep-pain relationship are not yet fully understood. Nevertheless, several studies already highlighted the potential importance of several factors and their working mechanisms. In general, potential underlying mechanisms can be divided in two categories, i.e., psychological and physiological mechanisms.

Psychological mechanisms and factors of the sleep-pain relationship

Emotions, affect, mood, depression and anxiety

Both chronic pain and sleep have an effect on someone's mood (including both positive and negative affect states [PA/NA]) and vice versa (29, 30). Higher NA potentially increases arousal and hypervigilance to pain, causing sensitisation to pain, avoidance, and functional disability (31). One study in a heterogeneous chronic pain population found that negative mood mediates the relationship between pain and sleep (32). Also, PA seems to attenuate both the perception of pain and the negative affective response to pain, increasing the resilience of the individual (33). Additionally, the absence of PA is related to poor pain-related outcomes (33). Furthermore, sleep disturbance appears to indirectly predict pain interference via NA and PA (34). Both NA and PA also seem to mediate the relationship between total sleep time and pain interference (34). In fibromyalgia patients, PA was found to mediate the relation between sleep quality and activity interference, sleep quality predicted lower levels of PA in the morning and PA predicted elevated end-of-day activity interference (35). Another study in people with rheumatoid arthritis and fibromyalgia showed that sleep duration moderates the association of pain and PA (36). While PA and NA are the opposite of each other, they represent different constructs that seem to contribute independently to the variance of the sleep-pain association (15, 33). Furthermore, PA appears to attenuate the relation of NA and chronic pain.

Depression is a prevalent comorbidity in both people with insomnia and people with chronic pain (37, 38). The presence of insomnia, chronic pain and depression or anxiety has been described before as the unhappy triad (37). Depressive symptoms partially mediate the relationship between sleep impairment and chronic pain development (39, 40). In fibromyalgia patients, sleep quality appears to mediate the relationship between pain and symptoms of depression (41). Another study in fibromyalgia patients indicated that pain was a mediator of the pathway from sleep impairment to depressive symptoms (36). Additionally, a community-based study observed that people with persistent pain who had a new period of sleep problems three years later had more than trebled the risk of a new period of probable depression at six years (42). Despite depressive symptoms being closely related to both sleep and pain, it appears that sleep and pain can vary beyond the influence of depression (15, 38). Another psychosocial factor that appears to play a crucial role in the sleep-pain relationship is anxiety. Anxiety symptoms appear to partially mediate the relationship between insomnia symptoms and pain incidence (39, 40). Overall, NA, PA, depression and anxiety can be essential contributing factors which can be potential targets for future treatment strategies. Nevertheless, there is a need for studies investigating the interrelationships between sleep, pain, and psychological outcomes and the potential effect of treatment strategies targeting psychosocial factors in people with chronic pain and sleep problems.

Pain-related beliefs, catastrophising and coping

Pain catastrophising is a maladaptive cognitive coping style (which entails the feeling of helplessness, rumination [i.e., inability to reduce pain-related thoughts] and magnification [i.e., tendency to exaggerate the threat level of a painful stimulus]) in response to anticipated or actual pain (43). While catastrophising is generally associated with NA, it is very different from clinical anxiety or depression. While pain catastrophising increases the risk of sleep disturbances in patients with chronic pain, people with both chronic pain and sleep problems also seem to pain catastrophise more (44). Furthermore, it indirectly influences pain severity and interference through sleep disturbance (45).

Catastrophising and emotional distress also mediate the relationship between sleep disturbance and chronic pain intensity (46). In people with temporomandibular disorder, pain catastrophising was associated with greater sleep disturbance, and sleep disturbances appear to mediate clinical pain severity and pain-related interference attributed to pain catastrophising (45). Knee osteoarthritis patients with lower sleep efficiency and greater catastrophising also had increased levels of central sensitisation (47). Another study in chronic non-malignant pain patients indicated that pain was a mediator in the relationship between sleep and pain catastrophising (48). It is also suggested that the potential inability to silence intrusive pain-related thoughts prior to bedtime in people who catastrophise a lot may lead to overall worse sleep (49). Furthermore, pain-catastrophising and pain-related thoughts and beliefs frequently impact an individual's daytime functioning and behaviour. This often leads to a less active lifestyle in chronic pain patients (50). As engagement in physical activity has a sleep-promoting effect (by increasing sleep pressure), the change to a less active lifestyle might negatively impact sleep. In addition to avoidance behaviour of some chronic pain patients, people with chronic pain and sleep problems often adopt coping strategies to compensate for their sleep problems (such as taking naps, staying longer in bed, and performing tasks in the bed room) (51, 52). These adopted strategies often have a negative influence on sleep in the long run since they reduce the sleep drive, and strengthen the association of arousal/sleepiness with the wrong environment (53). In summary, maladaptive beliefs, pain catastrophising and maladaptive coping strategies can interfere with both pain and sleep. Identifying and changing these maladaptive beliefs and behaviours are core features of pain management programmes that build on cognitive-behavioural principles.

Expectation and placebo/nocebo effects during sleep

Several studies show that expectation of pain plays a role in cortical arousal. In contrast, a reduction in cortical arousal (evoked by noxious stimuli during REM sleep) was found when analgesic expectations related to that stimulus were induced before sleep (54). Similarly, the induction of analgesic expectations before sleep led to reduced nocturnal pain perception, subjective sleep disturbances, and activated brain processes that modulate incoming nociceptive signals differentially according to sleep stage (55). REM sleep was also identified as a moderator of the relationship between pain relief expectations and placebo analgesia (56). Despite findings that expectations of higher pain levels can lead to higher pain sensitivity, expectations of higher pain levels did not appear to explain sleep restriction-related hyperalgesia, which suggests that they are mediated by different cortical mechanisms (57).

Physiological mechanisms and factors of the sleep-pain relationship

Endogenous pain modulation (EPM)

EPM includes all endogenous mechanisms that affect nociceptive signal processing (i.e., pain facilitation and inhibition mechanisms), which are controlled by the central nervous system (58). Both increased pain facilitation and impaired pain inhibition are signs of deficient EPM. Sleep influences EPM through several ways. Sleep loss often results in an elevation of pain signals by increasing pain facilitation and reduction of pain inhibition, habituation and tolerance (20, 59–61). These changes in central modulatory pain pathways are associated with a higher risk of chronic pain, highlighting that sleep impairment can contribute to the development of chronic pain through central pathways (62–64). In people with insomnia, both pain facilitation and inhibition appear to be altered but pain inhibition seems to be more strongly affected by sleep disturbances

(65). Additionally, it appears that pain-inhibitory circuits in insomniacs are in a state of constant activation to compensate for ongoing subclinical pain (65). Also, changes in pain habituation may appear only after chronic sleep deprivation and are most-likely reverted after a prolonged sleep recovery period (60). The impact of sleep impairment (i.e., insomnia or induced sleep deprivation) on EPM also appears to be dependent on sex, with females being more susceptible (15, 59). For example, pain inhibition seems to be mainly reduced in females (59). Also, secondary hyperalgesia is mainly seen in males, whereas sleep disruption increased temporal summation in females (61). A study using a three-week partial sleep deprivation protocol (with limited recovery) in healthy people indicated that exposure to chronic insufficient sleep has the potential to increase vulnerability to chronic pain by altering processes of pain habituation and sensitisation (60). Specifically in people with low back pain and fibromyalgia, lower subjective sleep quality was related to decreased pain inhibition (66, 67). Furthermore, similar finding of altered EPM have been found in other pain populations (68, 69). In general, changes in EPM are likely an important underlying mechanism of the sleep-pain relationship. However, longitudinal clinical studies are necessary to further establish the importance of central pain modulatory systems in the sleep-pain relationship and the long-term temporal association.

Opioidergic systems

Opioid peptides appear to play an important mediating role in descending inhibitory pain modulatory systems (70, 71). In several chronic pain conditions, limited pain-inhibitory capacity with reduced μ-opioid activation has been observed (72–74). Studies show that sleep deprivation dysregulates endogenous pain inhibition processes (including endogenous opioid systems), attenuates the analgesic efficacy of μ-opioid receptor agonists and induces hyperalgesia (20, 75, 76). Opioid receptors are located in multiple nuclei that actively regulate both sleep and pain (77). Also, animal studies highlighted that sleep deprivation alters μ- and δ-opioid receptor function in mesolimbic circuits (78), diminishes basal endogenous opioid levels (79), and downregulates central opioid receptors (78). Overall, the current literature suggests that endogenous opioid systems likely have a mediating role in the sleep-pain relationship.

Monoaminergic systems: dopamine, serotonin and norepinephrine

Dopamine is a well-known neurotransmitter, which has a key role in the reward motivation system, movement control and sleep-wake cycle. Regarding sleep, higher levels of dopamine are associated with more wake time while less dopamine is rather associated with sleepiness (80, 81). Previous animal studies observed an association between acute sleep deprivation and downregulated D2/3 receptor activity with lower receptor availability (82). This could partially explain the reduced alertness after sleep deprivation (82). Also, it is likely that the down-regulation of D2/3 receptors activity affects other systems where dopamine is involved, such as the reward motivation and pain modulation systems (19, 77, 83). Chronic pain could interfere with motivation by impairment of the mesolimbic dopaminergic system (80). Furthermore, dopamine is also related with antinociceptive effects and motivated behaviour in chronic pain patients (83, 84). It seems like lower levels of dopamine decrease protection towards pain, which consequently leads to facilitation of nociception (83). Such a dysfunction of the dopaminergic system has already been observed in people with fibromyalgia (85). Last, several genotypes related to molecules involved in dopamine neurotransmission could be associated with lower PA in the presence of greater pain in fibromyalgia patients (86, 87). This suggests that changes in dopamine signalling because of sleep disturbance could have a negative impact on pain processes.

The serotonergic system plays an important role in the mu-opioid analgesic function involved in EPM (68). It appears that serotonin has dual action with proalgesic effects in the periphery and pro- and/or analgesic effects in the CNS (88, 89). Serotonin also promotes wakefulness and inhibits REM sleep (90). Also deep non-REM sleep increases with inhibition of serotonin (90). Acute sleep deprivation results in an increase of serotonin plasma metabolites which possibly mediate the anti-depressive effect of acute sleep deprivation (91). Additionally, in animal studies, extracellular serotonin metabolites increase with acute sleep deprivation and the sensitivity of serotonin 1A receptors reduces with more chronic sleep restriction (92, 93).

Norepinephrine plays an important role in promoting wakefulness and arousal (94). With sleep onset, the norepinephrine levels in the blood circulation decline and norepinephrine levels are clearly lower during sleep compared to wakefulness (93). With sleep deprivation, the norepinephrine blood levels increase (95). Also, it is suggested that norepinephrine and the enhancements of norepinephrine transmission have an analgesic effect which synergises (i.e., becomes stronger) with when both noradrenergic and serotonergic transmissions are enhanced (95). It seems likely that the noradrenergic system mainly has an analgesic effect and rather limited influence on the pain-promoting effects of sleep deprivation (95).

Although it is likely that dopamine, serotonin and epinephrine are important factors in the relation between sleep and pain, further research is necessary to clarify how sleep deprivation alters the monoaminergic systems and to investigate whether these changes are correlated with concurrent changes in self-reported and measured pain sensitivity.

Inflammation

Several studies show that both sleep and pain can influence and be affected by inflammation, and that proinflammatory cytokines, such as IL-1β, IL-6, TNF-α and prostaglandins, are involved (96–98). Chronic sleep deprivation can lead to a low-grade inflammatory state with increased levels of proinflammatory markers (99, 100). This low-grade inflammation can be induced through different mechanisms, such as intestinal dysbiosis, impairment of HPA axes, circadian sleep disruption, obesity, or physical inactivity among others (17, 97). Furthermore, such inflammatory profiles can mediate pain responses (18). Based on preclinical observations, it has been suggested that glial overactivity, as seen in patients with chronic pain, is triggered by poor sleep (101). While glial activation leads to the production of inflammatory markers to favour tissue healing and restore homeostasis, chronic activation of glia cells can become pathogenic and can lead to collateral damage of nearby neurons and other glia (102). It is suggested that such glia overactivity can lead to central sensitisation and the development of chronic pain (101). Brain glia activation has already been found in people with chronic low back pain (103, 104), migraine (105) and fibromyalgia (106). Overall, it is likely that inflammation plays a role in the sleep-pain relationship considering that sleep problems can lead to changes in the immune response which can negatively impact the course of chronic pain (17, 19, 95). Furthermore, the literature suggests that chronic pain disorders can induce a low-grade inflammation, which can dysregulate sleep (15, 17, 95, 101, 107). Nevertheless, future research could further investigate the causality of the associations and whether improvement of sleep in people with chronic pain can reduce the inflammatory state and can reverse the negative consequences of sleep impairment.

The role of other endogenous substances

Several other substances seem to play a role in both pain modulation and sleep-wake regulation (17, 95). Adenosine, for example, has both proalgesic and analgesic effects and also has sleep-promoting properties (95, 108). Another example is melatonin, which is involved not only in circadian rhythms but also in physiological functions (such as foetal development, temperature regulation, cardiovascular regulation and immune system regulation) and pain regulation (i.e., melatonin can downregulate inflammatory mediators and has analgesic properties by its interaction with endorphins, GABA receptor, opioid receptors and the nitric oxide-arginine pathway) (17, 95). Melatonin is well known for its sleep-promoting effects but it also interacts with different pathways involved in pain (17). Furthermore, it is suggested that melatonin could have a positive effect on pain through circadian rhythm normalisation and its interaction with melatonin receptors and several neurotransmitter systems (109). Results of animal studies also show that suppressed melatonin secretion as result of sleep deprivation can increase glial activation and aggravate neuropathic pain (110). Beside adenosine and melatonin, other molecules that seem to influence both sleep and pain are orexin, nitric oxide and vitamin D (15, 17, 95). However, most endogenous mechanisms are still not fully understood.

Management of sleep problems in chronic pain and future research avenues

The shift from a rather biomedical pain model towards a biopsychosocial model in the management of chronic pain also led to the introduction of several treatment approaches targeting lifestyle factors and other psychosocial factors (111). Currently, the recommended treatment for chronic pain is a personalised multimodal, interdisciplinary approach, which aims to give the patient the required tools to manage their pain in the long term (7). Depending on the need of the patient, the following aspects are considered key components of the non-pharmacological chronic pain management: [1] pain neuroscience education to address maladaptive beliefs, [2] sleep management, [3] stress management, [4] nutrition care and [5] physical activity program (111–116). However, despite the increasing literature highlighting the impact of sleep impairment in people with chronic pain, the importance of sleep is rarely recognised in clinical practice or clinicians have limited knowledge of how to address sleep problems properly (117). If sleep problems are addressed, the treatment is often limited to the prescription of sedative pain/ sleep medications and/or education of sleep hygiene rules (13). Nevertheless, research investigating the implementation of treatment strategies to improve sleep shows promising results in several chronic pain populations.

One of the most investigated strategies is the use of cognitive behavioural therapy for insomnia (CBT-I). Currently, CBT-I is the best-evidence non-invasive, non-pharmacological treatment to address insomnia (118–120). Usually, CBT-I consists of several components including general sleep education, cognitive therapy, stimulus control, sleep restriction, sleep hygiene and relaxation therapy. However, variations in CBT-I programmes exist, usually with a selection of several of the above-mentioned components (120). A recent systematic review and meta-analysis indicates that the use of CBT-I programmes and hybrid treatment programmes in chronic pain populations overall results in improvements in sleep and pain parameters (121). However, the positive effect of CBT-I on pain seems rather moderate and short-term (121). This suggests that, while CBT-I is very effective in people with insomnia, some adaptations to CBT-I, taking into account the sleep–pain relationship and underlying mechanisms, are most likely necessary to optimise the treatment (121). Furthermore, there are limited to no significant effects on other outcomes (such as anxiety symptoms and fatigue) (121). Despite the promising initial findings of CBT-I, further research is necessary to gain more insight in the sleep-pain interaction and underlying mechanisms, and to continue to explore ways to enhance the treatment effects and efficiency.

There is still a lot of uncertainty about the mechanisms explaining sleep problems in chronic pain and their underlying interaction. However, there is no doubt that there is an association between sleep problems and pain, both on physiological and psychological levels. It seems warranted to have a better understanding of these mechanisms to be able to properly implement new treatment strategies. Future research could also focus on the testing and implementing non-pharmacological treatment strategies to target common psychological problems in people with pain and sleep difficulties, such as depression, anxiety and pain catastrophising. Currently, considering the available evidence, the implementation of CBT-I in a comprehensive intervention programme appears to be the best way to address sleep problems in people with chronic pain. Furthermore, there is increasing attention for the development of a hybrid CBT approach that combines select components of CBT-I and CBT for pain management to improve treatment efficiency and treatment adherence (122, 123). It is suggested that synergistic benefits can be gained by adopting a hybrid CBT approach given the bidirectional relationship between sleep and pain (122). A feasibility study highlights the usability of hybrid CBT for pain-related insomnia and the satisfaction with such a programme (124). The study also revealed several practical suggestions for future implementations including shortening the duration of each treatment session, reducing the amount of assessment paperwork, and minimising the burden of sleep and pain monitoring (124).

Additionally, it is likely that there are different endotypes (i.e., subtype of a disease condition) of the sleep–pain interaction, which present different pathways and mechanisms (125). Consequently, more specific management strategies could be selected based on the underlying pathways and mechanisms, which can lead to a more personalised approach (125). Therefore, the identification of different phenotypes (i.e., observable characteristics, traits or clinical presentation without implication of the mechanism) in individuals with sleep and pain problems could be helpful to discover different endotypes (125). While such adapted hybrid approaches show promise in optimising treatment effectiveness and flexibility, further research is required to identify phenotypes and ultimately endotypes, and to establish clinical and cost-effectiveness of hybrid CBT approaches.

References

1 Raja SN, Carr DB, Cohen M, Finnerup NB, Flor H, Gibson S, et al. The revised International Association for the Study of Pain definition of pain: concepts, challenges, and compromises. Pain. 2020;161(9):1976–82.

2 Clauw DJ, Essex MN, Pitman V, Jones KD. Reframing chronic pain as a disease, not a symptom: rationale and implications for pain management. Postgrad Med. 2019;131(3):185–98.

3 The IASP Taxonomy Working Group. Classification of Chronic Pain. Descriptions of Chronic Pain Syndromes and Definitions of Pain Terms (Revised). 2nd ed. Washington, DC: IASP Press; 2011.

4 Levinthal DJ, Bielefeldt K. Pain without nociception? Eur J Gastroenterol Hepatol. 2012;24(3):336–9.

5 Nijs J, Paul van Wilgen C, Van Oosterwijck J, van Ittersum M, Meeus M. How to explain central sensitization to patients with 'unexplained' chronic musculoskeletal pain: practice guidelines. Man Ther. 2011;16(5):413–18.

6 Dueñas M, Ojeda B, Salazar A, Mico JA, Failde I. A review of chronic pain impact on patients, their social environment and the health care system. J Pain Res. 2016;9:457–67.

7 Cohen SP, Vase L, Hooten WM. Chronic pain: an update on burden, best practices, and new advances. Lancet. 2021;397(10289):2082–97.

8 St Sauver JL, Warner DO, Yawn BP, Jacobson DJ, McGree ME, Pankratz JJ, et al. Why patients visit their doctors: assessing the most prevalent conditions in a defined American population. Mayo Clin Proc. 2013;88(1):56–67.

9 GBD 2019 Diseases and Injuries Collaborators. Global burden of 369 diseases and injuries in 204 countries and territories, 1990–2019: a systematic analysis for the Global Burden of Disease Study 2019. Lancet. 2020;396(10258):1204–22.

10 Dahlhamer J, Lucas J, Zelaya C, Nahin R, Mackey S, DeBar L, et al. Prevalence of chronic pain and high-impact chronic pain among adults – United States, 2016. MMWR Morb Mortal Wkly Rep. 2018;67(36):1001–6.

11 Gatchel RJ, Peng YB, Peters ML, Fuchs PN, Turk DC. The biopsychosocial approach to chronic pain: scientific advances and future directions. Psychol Bull. 2007;133(4):581–624.

12 Meints SM, Edwards RR. Evaluating psychosocial contributions to chronic pain outcomes. Prog Neuropsychopharmacol Biol Psychiatry. 2018;87(Pt B):168–82.

13 Tang NK, Wright KJ, Salkovskis PM. Prevalence and correlates of clinical insomnia co-occurring with chronic back pain. J Sleep Res. 2007;16(1):85–95.

14 Sun Y, Laksono I, Selvanathan J, Saripella A, Nagappa M, Pham C, et al. Prevalence of sleep disturbances in patients with chronic non-cancer pain: a systematic review and meta-analysis. Sleep Med Rev. 2021;57:101467.

15 Finan PH, Goodin BR, Smith MT. The association of sleep and pain: an update and a path forward. J Pain. 2013;14(12):1539–52.

16 Van Looveren E, Bilterys T, Munneke W, Cagnie B, Ickmans K, Mairesse O, et al. The association between sleep and chronic spinal pain: a systematic review from the last decade. J Clin Med. 2021;10(17):3836.

17 Herrero Babiloni A, De Koninck BP, Beetz G, De Beaumont L, Martel MO, Lavigne GJ. Sleep and pain: recent insights, mechanisms, and future directions in the investigation of this relationship. J Neural Transm (Vienna, Austria: 1996). 2020;127(4):647–60.

18 Afolalu EF, Ramlee F, Tang NKY. Effects of sleep changes on pain-related health outcomes in the general population: a systematic review of longitudinal studies with exploratory meta-analysis. Sleep Med Rev. 2018;39:82–97.

19 Andersen ML, Araujo P, Frange C, Tufik S. Sleep disturbance and pain: a tale of two common problems. Chest. 2018;154(5):1249–59.

20 Schuh-Hofer S, Wodarski R, Pfau DB, Caspani O, Magerl W, Kennedy JD, et al. One night of total sleep deprivation promotes a state of generalized hyperalgesia: a surrogate pain model to study the relationship of insomnia and pain. Pain. 2013;154(9):1613–21.

21 Haack M, Sanchez E, Mullington JM. Elevated inflammatory markers in response to prolonged sleep restriction are associated with increased pain experience in healthy volunteers. Sleep. 2007;30(9):1145–52.

22 Matre D, Hu L, Viken LA, Hjelle IB, Wigemyr M, Knardahl S, et al. Experimental sleep restriction facilitates pain and electrically induced cortical responses. Sleep. 2015;38(10):1607–17.

23 Gupta A, Silman AJ, Ray D, Morriss R, Dickens C, MacFarlane GJ, et al. The role of psychosocial factors in predicting the onset of chronic widespread pain:

results from a prospective population-based study. Rheumatology (Oxford). 2007;46(4):666–71.

24 Ostovar-Kermani T, Arnaud D, Almaguer A, Garcia I, Gonzalez S, Mendez Martinez YH, et al. Painful sleep: insomnia in patients with chronic pain syndrome and its consequences. Folia Med (Plovdiv). 2020;62(4):645–54.

25 Tang NK, Goodchild CE, Sanborn AN, Howard J, Salkovskis PM. Deciphering the temporal link between pain and sleep in a heterogeneous chronic pain patient sample: a multilevel daily process study. Sleep. 2012;35(5):675–87a.

26 Tang NK, Sanborn AN. Better quality sleep promotes daytime physical activity in patients with chronic pain? A multilevel analysis of the within-person relationship. PLoS ONE. 2014;9(3):e92158.

27 Bilterys T, Siffain C, De Maeyer I, Van Looveren E, Mairesse O, Nijs J, et al. Associates of insomnia in people with chronic spinal pain: a systematic review and meta-analysis. J Clin Med. 2021;10(14).

28 Jungquist CR, O'Brien C, Matteson-Rusby S, Smith MT, Pigeon WR, Xia Y, et al. The efficacy of cognitive-behavioral therapy for insomnia in patients with chronic pain. Sleep Med. 2010;11(3):302–9.

29 Finan PH, Quartana PJ, Remeniuk B, Garland EL, Rhudy JL, Hand M, et al. Partial sleep deprivation attenuates the positive affective system: effects across multiple measurement modalities. Sleep. 2017;40(1).

30 Konjarski M, Murray G, Lee VV, Jackson ML. Reciprocal relationships between daily sleep and mood: a systematic review of naturalistic prospective studies. Sleep Med Rev. 2018;42:47–58.

31 Janssen SA. Negative affect and sensitization to pain. Scand J Psychol. 2002;43(2):131–7.

32 O'Brien EM, Waxenberg LB, Atchison JW, Gremillion HA, Staud RM, McCrae CS, et al. Intraindividual variability in daily sleep and pain ratings among chronic pain patients: bidirectional association and the role of negative mood. Clin J Pain. 2011;27(5):425–33.

33 Finan PH, Garland EL. The role of positive affect in pain and its treatment. Clin J Pain. 2015;31(2):177–87.

34 Ravyts SG, Dzierzewski JM, Raldiris T, Perez E. Sleep and pain interference in individuals with chronic pain in mid- to late-life: the influence of negative and positive affect. J Sleep Res. 2019;28(4):e12807.

35 Kothari DJ, Davis MC, Yeung EW, Tennen HA. Positive affect and pain: mediators of the within-day relation linking sleep quality to activity interference in fibromyalgia. Pain. 2015;156(3):540–6.

36 Hamilton NA, Pressman M, Lillis T, Atchley R, Karlson C, Stevens N. Evaluating evidence for the role of sleep in fibromyalgia: a test of the sleep and pain diathesis model. Cognit Ther Res. 2012;36(6):806–14.

37 Koffel E, Krebs EE, Arbisi PA, Erbes CR, Polusny MA. The unhappy triad: pain, sleep complaints, and internalizing symptoms. Clin Psychol Sci J Assoc Psychol Sci. 2016;4(1):96–106.

38 Wilson KG, Eriksson MY, D'Eon JL, Mikail SF, Emery PC. Major depression and insomnia in chronic pain. Clin J Pain. 2002;18(2):77–83.

39 Generaal E, Vogelzangs N, Penninx BW, Dekker J. Insomnia, sleep duration, depressive symptoms, and the onset of chronic multisite musculoskeletal pain. Sleep. 2017;40(1).

40 Dunietz GL, Swanson LM, Jansen EC, Chervin RD, O'Brien LM, Lisabeth LD, et al. Key insomnia symptoms and incident pain in older adults: direct and mediated pathways through depression and anxiety. Sleep. 2018;41(9).

41 Miró E, Martínez MP, Sánchez AI, Prados G, Medina A. When is pain related to emotional distress and daily functioning in fibromyalgia syndrome? The mediating roles of self-efficacy and sleep quality. Br J Health Psychol. 2011;16(4):799–814.

42 Campbell P, Tang N, McBeth J, Lewis M, Main CJ, Croft PR, et al. The role of sleep problems in the development of depression in those with persistent pain: a prospective cohort study. Sleep. 2013;36(11):1693–8.

43 Quartana PJ, Campbell CM, Edwards RR. Pain catastrophizing: a critical review. Expert Rev Neurother. 2009;9(5):745–58.

44 MacDonald S, Linton SJ, Jansson-Frojmark M. Avoidant safety behaviors and catastrophizing: shared cognitive-behavioral processes and consequences in co-morbid pain and sleep disorders. Int J Behav Med. 2008;15(3):201–10.

45 Buenaver LF, Quartana PJ, Grace EG, Sarlani E, Simango M, Edwards RR, et al. Evidence for indirect effects of pain catastrophizing on clinical pain among myofascial temporomandibular disorder participants: the mediating role of sleep disturbance. Pain. 2012;153(6):1159–66.

46 Burgess HJ, Burns JW, Buvanendran A, Gupta R, Chont M, Kennedy M, et al. Associations between sleep disturbance and chronic pain intensity and function: a test of direct and indirect pathways. Clin J Pain. 2019;35(7):569–76.

47 Campbell CM, Buenaver LF, Finan P, Bounds SC, Redding M, McCauley L, et al. Sleep, pain catastrophizing, and central sensitization in knee osteoarthritis patients with and without insomnia. Arthritis Care Res (Hoboken). 2015;67(10):1387–96.

48 Wilt JA, Davin S, Scheman J. A multilevel path model analysis of the relations between sleep, pain, and pain catastrophizing in chronic pain rehabilitation patients. Scand J Pain. 2016;10:122–9.

49 Smith MT, Perlis ML, Carmody TP, Smith MS, Giles DE. Presleep cognitions in patients with insomnia secondary to chronic pain. J Behav Med. 2001;24(1):93–114.

50 Larsson C, Ekvall Hansson E, Sundquist K, Jakobsson U. Impact of pain characteristics and fear-avoidance beliefs on physical activity levels among older adults with chronic pain: a population-based, longitudinal study. BMC Geriatr. 2016;16:50.

51 Råheim M, Håland W. Lived experience of chronic pain and fibromyalgia: women's stories from daily life. Qual Health Res. 2006;16(6):741–61.

52 Theadom A, Cropley M. 'This constant being woken up is the worst thing' – experiences of sleep in fibromyalgia syndrome. Disabil Rehabil. 2010;32(23):1939–47.

53 Van Looveren E, Meeus M, Cagnie B, Ickmans K, Bilterys T, Malfliet A, et al. Combining cognitive behavioral therapy for insomnia and chronic spinal pain within physical therapy: a practical guide for the implementation of an integrated approach. Phys Ther. 2022;102(8).

54 Laverdure-Dupont D, Rainville P, Renancio C, Montplaisir J, Lavigne G. Placebo analgesia persists during sleep: an experimental study. Prog Neuropsychopharmacol Biol Psychiatry. 2018;85:33–8.

55 Chouchou F, Dang-Vu TT, Rainville P, Lavigne G. The role of sleep in learning placebo effects. Int Rev Neurobiol. 2018;139:321–55.

56 Chouchou F, Chauny JM, Rainville P, Lavigne GJ. Selective REM sleep deprivation improves expectation-related placebo analgesia. PLoS ONE. 2015;10(12):e0144992.

57 Ree A, Nilsen KB, Knardahl S, Sand T, Matre D. Sleep restriction does not potentiate nocebo-induced changes in pain and cortical potentials. Eur J Pain. 2020;24(1):110–21.

58 Yarnitsky D. Role of endogenous pain modulation in chronic pain mechanisms and treatment. Pain. 2015;156(Suppl 1):S24–31.

59 Eichhorn N, Treede RD, Schuh-Hofer S. The role of sex in sleep deprivation related changes of nociception and conditioned pain modulation. Neuroscience. 2018;387:191–200.

60 Simpson NS, Scott-Sutherland J, Gautam S, Sethna N, Haack M. Chronic exposure to insufficient sleep alters processes of pain habituation and sensitization. Pain. 2018;159(1):33–40.

61 Smith MT, Jr., Remeniuk B, Finan PH, Speed TJ, Tompkins DA, Robinson M, et al. Sex differences in measures of central sensitization and pain sensitivity to experimental sleep disruption: implications for sex differences in chronic pain. Sleep. 2019;42(2).

62 Lewis GN, Rice DA, McNair PJ. Conditioned pain modulation in populations with chronic pain: a systematic review and meta-analysis. J Pain. 2012;13(10):936–44.

63 O'Brien AT, Deitos A, Triñanes Pego Y, Fregni F, Carrillo-de-la-Peña MT. Defective endogenous pain modulation in fibromyalgia: a meta-analysis of temporal summation and conditioned pain modulation paradigms. J Pain. 2018;19(8):819–36.

64 Staud R. Abnormal endogenous pain modulation is a shared characteristic of many chronic pain conditions. Expert Rev Neurother. 2012;12(5):577–85.

65 Haack M, Scott-Sutherland J, Santangelo G, Simpson NS, Sethna N, Mullington JM. Pain sensitivity and modulation in primary insomnia. Eur J Pain. 2012;16(4):522–33.

66 Klyne DM, Moseley GL, Sterling M, Barbe MF, Hodges PW. Individual variation in pain sensitivity and conditioned pain modulation in acute low back pain: effect of stimulus type, sleep, and psychological and lifestyle factors. J Pain. 2018;19(8):942. e1–18.

67 Paul-Savoie E, Marchand S, Morin M, Bourgault P, Brissette N, Rattanavong V, et al. Is the deficit in pain inhibition in fibromyalgia influenced by sleep impairments? Open Rheumatol J. 2012;6:296–302.

68 Lee YC, Lu B, Edwards RR, Wasan AD, Nassikas NJ, Clauw DJ, et al. The role of sleep problems in central pain processing in rheumatoid arthritis. Arthritis Rheum. 2013;65(1):59–68.

69 Edwards RR, Grace E, Peterson S, Klick B, Haythornthwaite JA, Smith MT. Sleep continuity and architecture: associations with pain-inhibitory processes in patients with temporomandibular joint disorder. Eur J Pain. 2009;13(10):1043–7.

70 Bencherif B, Fuchs PN, Sheth R, Dannals RF, Campbell JN, Frost JJ. Pain activation of human supraspinal opioid pathways as demonstrated by [11C]-carfentanil and positron emission tomography (PET). Pain. 2002;99(3):589–98.

71 Julien N, Marchand S. Endogenous pain inhibitory systems activated by spatial summation are opioid-mediated. Neurosci Lett. 2006;401(3):256–60.

72 Lautenbacher S, Rollman GB. Possible deficiencies of pain modulation in fibromyalgia. Clin J Pain. 1997;13(3):189–96.

73 Peters ML, Schmidt AJM, Van den Hout MA, Koopmans R, Sluijter ME. Chronic back pain, acute postoperative pain and the activation of diffuse noxious inhibitory controls (DNIC). Pain. 1992;50(2):177–87.

74 Martikainen IK, Peciña M, Love TM, Nuechterlein EB, Cummiford CM, Green CR, et al. Alterations in endogenous opioid functional measures in chronic back pain. J Neurosci. 2013;33(37):14729–37.

75 Nascimento DC, Andersen ML, Hipólide DC, Nobrega JN, Tufik S. Pain hypersensitivity induced by paradoxical sleep deprivation is not due to altered binding to brain mu-opioid receptors. Behav Brain Res. 2007;178(2):216–20.

76 Roehrs T, Hyde M, Blaisdell B, Greenwald M, Roth T. Sleep loss and REM sleep loss are hyperalgesic. Sleep. 2006;29(2):145–51.

77 Foo H, Mason P. Brainstem modulation of pain during sleep and waking. Sleep Med Rev. 2003;7(2):145–54.

78 Fadda P, Tortorella A, Fratta W. Sleep deprivation decreases mu and delta opioid receptor binding in the rat limbic system. Neurosci Lett. 1991;129(2):315–17.

79 Przewłocka B, Mogilnicka E, Lasón W, van Luijtelaar EL, Coenen AM. Deprivation of REM sleep in the rat and the opioid peptides beta-endorphin and dynorphin. Neurosci Lett. 1986;70(1):138–42.

80 Taylor AMW, Becker S, Schweinhardt P, Cahill C. Mesolimbic dopamine signaling in acute and chronic pain: implications for motivation, analgesia, and addiction. Pain. 2016;157(6):1194–8.

81 Monti JM, Jantos H. The roles of dopamine and serotonin, and of their receptors, in regulating sleep and waking. Prog Brain Res. 2008;172:625–46.

82 Volkow ND, Tomasi D, Wang GJ, Telang F, Fowler JS, Logan J, et al. Evidence that sleep deprivation downregulates dopamine D2R in ventral striatum in the human brain. J Neurosci. 2012;32(19):6711–17.

83 Finan PH, Remeniuk B. Is the brain reward system a mechanism of the association of sleep and pain? Pain Manag. 2016;6(1):5–8.

84 Schwartz N, Temkin P, Jurado S, Lim BK, Heifets BD, Polepalli JS, et al. Chronic pain. Decreased motivation during chronic pain requires long-term depression in the nucleus accumbens. Science. 2014;345(6196):535–42.

85 Wood PB, Patterson JC, Sunderland JJ, Tainter KH, Glabus MF, Lilien DL. Reduced presynaptic dopamine activity in fibromyalgia syndrome demonstrated with positron emission tomography: a pilot study. J Pain. 2007;8(1):51–8.

86 Finan PH, Zautra AJ, Davis MC, Lemery-Chalfant K, Covault J, Tennen H. Genetic influences on the dynamics of pain and affect in fibromyalgia. Health Psychol. 2010;29(2):134–42.

87 Finan PH, Zautra AJ, Davis MC, Lemery-Chalfant K, Covault J, Tennen H. COMT moderates the relation of daily maladaptive coping and pain in fibromyalgia. Pain. 2011;152(2):300–7.

88 Viguier F, Michot B, Hamon M, Bourgoin S. Multiple roles of serotonin in pain control mechanisms – implications of 5-HT$_7$ and other 5-HT receptor types. Eur J Pharmacol. 2013;716(1–3):8–16.

89 Steen KH, Steen AE, Reeh PW. A dominant role of acid pH in inflammatory excitation and sensitization of nociceptors in rat skin, in vitro. J Neurosci. 1995;15(5 Pt 2):3982–9.

90 Monti JM. Serotonin control of sleep-wake behavior. Sleep Med Rev. 2011;15(4):269–81.

91 Davies SK, Ang JE, Revell VL, Holmes B, Mann A, Robertson FP, et al. Effect of sleep deprivation on the human metabolome. Proc Natl Acad Sci U S A. 2014;111(29):10761–6.

92 Zant JC, Leenaars CH, Kostin A, Van Someren EJ, Porkka-Heiskanen T. Increases in extracellular serotonin and dopamine metabolite levels in the basal forebrain during sleep deprivation. Brain Res. 2011;1399:40–8.

93 Meerlo P, Sgoifo A, Suchecki D. Restricted and disrupted sleep: effects on autonomic function, neuroendocrine stress systems and stress responsivity. Sleep Med Rev. 2008;12(3):197–210.

94 Berridge CW, Schmeichel BE, España RA. Noradrenergic modulation of wakefulness/arousal. Sleep Med Rev. 2012;16(2):187–97.

95 Haack M, Simpson N, Sethna N, Kaur S, Mullington J. Sleep deficiency and chronic pain: potential underlying mechanisms and clinical implications. Neuropsychopharmacology. 2020;45(1):205–16.

96 Zhang JM, An J. Cytokines, inflammation, and pain. Int Anesthesiol Clin. 2007;45(2):27–37.

97 Mullington JM, Simpson NS, Meier-Ewert HK, Haack M. Sleep loss and inflammation. Best Pract Res Clin Endocrinol Metab. 2010;24(5):775–84.

98 Watkins LR, Milligan ED, Maier SF. Glial proinflammatory cytokines mediate exaggerated pain states: implications for clinical pain. Adv Exp Med Biol. 2003;521:1–21.

99 Simpson N, Dinges DF. Sleep and inflammation. Nutr Rev. 2007;65(12 Pt 2):S244–52.

100 Irwin MR, Olmstead R, Carroll JE. Sleep disturbance, sleep duration, and inflammation: a systematic review and meta-analysis of cohort studies and experimental sleep deprivation. Biol Psychiatry. 2016;80(1):40–52.

101 Nijs J, Clark J, Malfliet A, Ickmans K, Voogt L, Don S, et al. In the spine or in the brain? Recent advances in pain neuroscience applied in the intervention for low back pain. Clin Exp Rheumatol. 2017;107(5);35 Suppl:108–15.

102 Spielman LJ, Little JP, Klegeris A. Physical activity and exercise attenuate neuro-inflammation in neurological diseases. Brain Res Bull. 2016;125:19–29.

103 Loggia ML, Chonde DB, Akeju O, Arabasz G, Catana C, Edwards RR, et al. Evidence for brain glial activation in chronic pain patients. Brain. 2015;138(Pt 3):604–15.

104 Alshelh Z, Brusaferri L, Saha A, Morrissey E, Knight P, Kim M, et al. Neuroimmune signatures in chronic low back pain subtypes. Brain. 2022;145(3):1098–110.

105 Albrecht DS, Mainero C, Ichijo E, Ward N, Granziera C, Zürcher NR, et al. Imaging of neuroinflammation in migraine with aura: a [(11)C]PBR28 PET/MRI study. Neurology. 2019;92(17):e2038–50.

106 Albrecht DS, Forsberg A, Sandström A, Bergan C, Kadetoff D, Protsenko E, et al. Brain glial activation in fibromyalgia – a multi-site positron emission tomography investigation. Brain Behav Immun. 2019;75:72–83.

107 Selvanathan J, Tang NKY, Peng PWH, Chung F. Sleep and pain: relationship, mechanisms, and managing sleep disturbance in the chronic pain population. Int Anesthesiol Clin. 2022;60(2):27–34.

108 Besedovsky L, Lange T, Haack M. The sleep-immune crosstalk in health and disease. Physiol Rev. 2019;99(3):1325–80.

109 Danilov A, Kurganova J. Melatonin in chronic pain syndromes. Pain Ther. 2016;5(1):1–17.

110 Huang CT, Chiang RP, Chen CL, Tsai YJ. Sleep deprivation aggravates median nerve injury-induced neuropathic pain and enhances microglial activation by suppressing melatonin secretion. Sleep. 2014;37(9):1513–23.

111 Malfliet A, Ickmans K, Huysmans E, Coppieters I, Willaert W, Bogaert WV, et al. Best evidence rehabilitation for chronic pain part 3: low back pain. J Clin Med. 2019;8(7).

112 Hylands-White N, Duarte RV, Raphael JH. An overview of treatment approaches for chronic pain management. Rheumatol Int. 2017;37(1):29–42.

113 Nijs J, Mairesse O, Neu D, Leysen L, Danneels L, Cagnie B, et al. Sleep disturbances in chronic pain: neurobiology, assessment, and treatment in physical therapist practice. Phys Ther. 2018;98(5):325–35.

114 Louw A, Sluka KA, Nijs J, Courtney CA, Zimney K. Revisiting the provision of pain neuroscience education: an adjunct intervention for patients but a primary focus of clinician education. J Orthop Sports Phys Ther. 2021;51(2):57–9.

115 Elma Ö, Yilmaz ST, Deliens T, Clarys P, Nijs J, Coppieters I, et al. Chronic musculoskeletal pain and nutrition: where are we and where are we heading? PM & R J Inj Funct Rehab. 2020;12(12):1268–78.

116 Nijs J, Tumkaya Yilmaz S, Elma Ö, Tatta J, Mullie P, Vanderweeën L, et al. Nutritional intervention in chronic pain: an innovative way of targeting central nervous system sensitization? Expert Opin Ther Targets. 2020;24(8):793–803.

117 Ogeil RP, Chakraborty SP, Young AC, Lubman DI. Clinician and patient barriers to the recognition of insomnia in family practice: a narrative summary of reported literature analysed using the theoretical domains framework. BMC Fam Pract. 2020;21(1):1.

118 Trauer JM, Qian MY, Doyle JS, Rajaratnam SM, Cunnington D. Cognitive behavioral therapy for chronic insomnia: a systematic review and meta-analysis. Ann Intern Med. 2015;163(3):191–204.

119 van der Zweerde T, Bisdounis L, Kyle SD, Lancee J, van Straten A. Cognitive behavioral therapy for insomnia: a meta-analysis of long-term effects in controlled studies. Sleep Med Rev. 2019;48:101208.

120 van Straten A, van der Zweerde T, Kleiboer A, Cuijpers P, Morin CM, Lancee J. Cognitive and behavioral therapies in the treatment of insomnia: a meta-analysis. Sleep Med Rev. 2018;38:3–16.

121 Selvanathan J, Pham C, Nagappa M, Peng PWH, Englesakis M, Espie CA, et al. Cognitive behavioral therapy for insomnia in patients with chronic pain – a systematic review and meta-analysis of randomized controlled trials. Sleep Med Rev. 2021;60:101460.

122 Tang NKY. CBT-I protocol for insomnia co-morbid with chronic pain. In: Baglioni, C., Espie, C.A., Riemann, D., editors. Cognitive-Behavioural Therapy for Insomnia (CBT-I) Across the Life Span; Wiley-Blackwell; 2022. p. 169–79.

123 Tang NKY. Cognitive behavioural therapy in pain and psychological disorders: towards a hybrid future. Prog Neuropsychopharmacol Biol Psychiatry. 2018;87(Pt B):281–9.

124 Tang NKY, Moore C, Parsons H, Sandhu HK, Patel S, Ellard DR, et al. Implementing a hybrid cognitive-behavioural therapy for pain-related insomnia in primary care: lessons learnt from a mixed-methods feasibility study. BMJ Open. 2020;10(3):e034764.

125 Herrero Babiloni A, Beetz G, Tang NKY, Heinzer R, Nijs J, Martel MO, et al. Towards the endotyping of the sleep-pain interaction: a topical review on multi-target strategies based on phenotypic vulnerabilities and putative pathways. Pain. 2021;162(5):1281–8.

10

SLEEP, CIRCADIAN RHYTHMS AND SHIFT WORK

Adopting personalised approaches to managing shift work and circadian misalignment

Lin Shen, Prerna Varma, Jade M. Murray and Tracey L. Sletten

Circadian rhythms

Circadian rhythms are endogenously generated, daily cycles in physiological and behavioural processes (1), including core body temperature, alertness and sleep/wake timings (2, 3). These rhythms are generated by the periodic expression of clock genes in the central circadian pacemaker located in the suprachiasmatic nuclei (SCN) of the hypothalamus (4), in addition to peripheral clocks located in many cells and tissues throughout the body (5, 6). The SCN controls the circadian rhythm in the production and secretion of the hormone melatonin, with production increasing in the biological evening and reaching its peak during the biological night, before decreasing again in the biological morning (7, 8). In humans, melatonin has a sleep-promoting effect and is closely linked to the behavioural sleep-wake cycle in healthy, nocturnal sleepers. Sleep is initiated as melatonin levels increase at night, is maintained via the stable melatonin levels during the biological night, and wakefulness is promoted in the biological morning as melatonin levels decrease (9, 10).

Although circadian rhythms are generated in the absence of external time cues, the intrinsic circadian period averages 24.2 hours (1) and, therefore, requires input from environmental time cues, or zeitgebers, to maintain synchrony with the 24-hour environment (11). This synchrony is achieved through changes in the timing of the circadian pacemaker to an earlier or later time (i.e., circadian phase advancing or delaying, respectively) (12). Circadian phase refers to a timepoint within the circadian cycle and can be measured in terms of the peak or nadir of the rhythm (13), or based on other points, such as the onset of hormone synthesis (e.g., melatonin) (14). Circadian phase

DOI: 10.4324/9781003296966-10

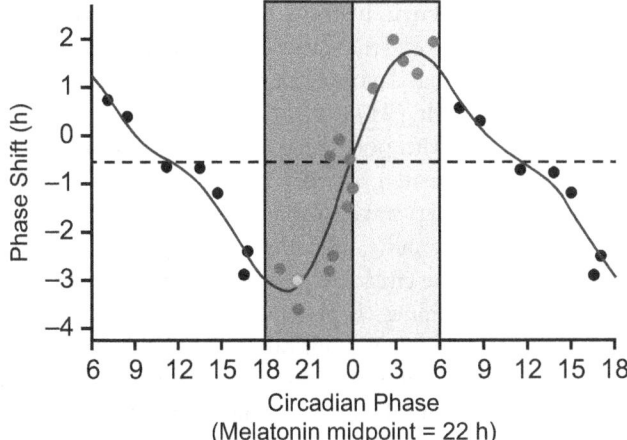

FIGURE 10.1 The human phase response curve to light. The magnitude of phase shift (shown on the y-axis) is dependent on the circadian timing of light exposure. The x-axis shows healthy subjects' circadian phase, with 0 representing minimum core body temperature (CBTmin). As indicated by the bars on either side of 0, light exposure approximately 3 hours before and after CBTmin induces the greatest phase delays and advances respectively. Figure from Stone and colleagues (15), with permission.

shifting induced by zeitgebers is assessed by measuring changes in circadian phase across multiple endogenous cycles (14).

The light-dark cycle is the primary zeitgeber in humans (16). The effects of light on circadian rhythms can be described by the human Phase Response Curve (PRC; Figure 10.1) (17). Core body temperature minimum (CBTmin) typically occurs 2–3 hours before the end of nocturnal sleep in day-working, healthy sleepers (18), and is represented by 0 on the abscissa in Figure 10.1. Light presented before the CBTmin induces phase delays (i.e., dark grey bar in Figure 10.1), whereas light presented after the CBTmin induces phase advances (17) (i.e., light grey bar in Figure 10.1). Considerable disturbances occur, however, when the timing of the circadian pacemaker becomes misaligned with the timing of the sleep-wake cycle, such as in individuals undertaking shift work, travelling across time zones, and in conditions such as delayed sleep-wake phase disorder. By far, the most common cause of circadian misalignment is shift work.

Shift work

The prevalence of shift work has increased over recent decades, driven by an accelerated demand for round-the-clock goods and services. An estimated 15–30% of the global workforce engages in shift work (19, 20). Shift work characteristics differ between industries, occupations and organisations but can be broadly defined as 'work [that] usually encompasses work time arrangements outside of conventional daytime hours, which includes fixed early morning, evening, and night work' (21).

Shift work is associated with poor sleep and circadian disruption due to circadian misalignment, a mismatch between endogenous rhythms and timing of daily behaviours such as sleep-wake, feeding-fasting and light exposure cycles (22). In the context of night shifts, individuals are expected to be awake during the biological night when the circadian pacemaker is promoting sleep and sleep during the day when wakefulness is promoted (23). Sleeping at inappropriate phases of the circadian cycle, such as during the daytime when melatonin levels are low, is associated with shorter and less consolidated sleep (including longer time to fall asleep and more awakenings during sleep) (24). Chronic sleep deprivation is well documented in shift workers (25) and tends to be particularly restricted for daytime sleep obtained between consecutive night shifts (24).

Working non-standard hours, such as in shift work, is associated with adverse consequences for health and well-being, productivity and safety (21), including mental health conditions (26), cardiovascular disorders (27), and cancers (28). Working during the biological night when the circadian pacemaker is promoting sleep is linked to decreased alertness and performance, increased rates of errors, workplace accidents and injuries (29). These safety risks extend beyond the workplace with shift work identified as the greatest sleep-related factor contributing to motor vehicle crashes (30). The commute home after night shift, in particular, is well associated with impaired alertness and adverse driving events (31), especially when the commute occurs close to the peak levels of melatonin (32). Alertness impairments may be particularly worse on the first shift in a sequence of night shifts as workers also often report extended wakefulness prior to the first night shift (24).

Individual differences in tolerance and circadian response to shift work

While all shift workers are at risk of adverse sleep and alertness outcomes, there are individual variations in shift work tolerance, defined as the ability to adapt to shift work with minimal impact on sleep, fatigue and functional impairments (33, 34). Depending on the method of assessment, between 20 and 44% of shift workers experience shift work disorder (35, 36), characterised by chronic and excessive sleepiness and insomnia that require extended recovery periods (37).

A complex interplay of biological, psychosocial and organisational factors contributes to these individual differences. For instance, diurnal preference, an individual's preferred timings for sleep/wake and daily activities, predicts tolerance to different shifts. Morning preference predicts better adaptation to day shifts and poorer adaptation to night shifts, whereas evening preference predicts poorer adaptation to day shifts (38). An individual's genetic makeup is also an important predictor of shift work tolerance. As one example, individuals with the long polymorphism in the PERIOD3 gene (PER3$^{-/5}$) have greater morning preference (39) and poorer tolerance for night shift work (40). Compared to male shift workers, females report more sleep problems (41), higher fatigue and poorer alertness (42). However, they also report better mental health (43) and record less cognitive impairment (44). Younger shift workers record lower fatigue on night shifts and lower risk for shift work disorder (45, 46). That being said, older age and more shift work experience may buffer against shift work-related consequences, with older age predicting better alertness on night shifts and lower absenteeism from work (47, 48).

Large variability in the response of the circadian system to shift work has also been consistently reported (15, 49). A recent study found a 4.5-hour range in circadian timing when rotating shift workers were on a day schedule, as well as considerable differences in the magnitude and direction of circadian phase shift in response to the same shift rotation (Figure 10.2) (15). When measured after 3–4 consecutive night shifts, there was a 6.8-hour range in participants' circadian timing, ranging from a 3:43-hour delay to 3:07-hour advance in circadian timing compared to day schedule (15). Individuals' patterns of light exposure along with diurnal preference were major determinants of the circadian response

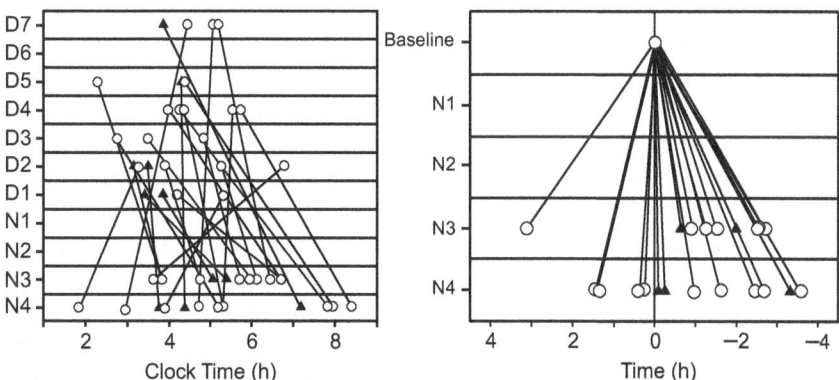

FIGURE 10.2 Variability in participants' circadian phase shift across a rotating shift schedule. Left panel: clock time of participants' circadian timing on a day schedule (D1-D7) and on final night shift (N3 or N4). Right panel: change in participants' circadian timing from baseline (i.e., day schedule) to the final night shift. Men are represented by closed triangles and women by open circles. Figure from Stone and colleagues (15), with permission.

to night shift, with more light in the three hours before/after the CBTmin associated with greater phase advance/delay (Figure 10.1) (15). More recent evidence also demonstrate substantial differences in individuals' sensitivity to evening light (50), which may explain individual differences in vulnerability to shift work and the circadian adaptation to shift work.

Interventions

Managing shift work and improving sleep

The plethora of shift work-related consequences highlights a strong need for interventions to support shift workers. However, the extent of individual differences and the added complexity of variability in shift schedules make it challenging to implement strategies in a shift working context. Interventions can occur at an organisational or individual level; existing organisational strategies typically use a 'one-size-fits-all' approach, whereas individual strategies are grounded in sleep hygiene and psychoeducation.

Organisational level strategies include rostering approaches to consider multiple interacting factors that facilitate sufficient opportunity to sleep between shifts. Factors such as the length and timing of shifts, duration of break between shifts, speed and direction of rotation between different shift types (i.e., forward vs backward), all contribute to health and safety outcomes (51). Generally, forward rotating shift schedules (i.e., day-evening-night shifts) allow for better adaptation to work, typically providing greater sleep opportunity and aligning with the circadian pacemaker's natural tendency to delay (21). The duration of breaks between shifts is also important as shorter break duration decreases sleep opportunity (51). Rostering practices need to also consider the role of time of day. Breaks occurring in the evening or night-time allow for nocturnal sleep, and therefore result in longer and higher quality sleep (52).

Individual-level strategies focus primarily on behavioural countermeasures, including napping, caffeine use and strategic light exposure to improve alertness on shift, and/or improve circadian adaptation to shift work. The efficacy and desirability of these countermeasures depend on individuals' shift patterns. Improving circadian adaptation under rapidly rotating rosters or isolated shifts is not recommended as the circadian system cannot adapt quickly enough to provide benefit (53), and will also impair individuals' sleep and functioning on the days prior to the isolated shift (for example), and prolong re-adaptation to a day schedule after the shift. Instead, strategies can be implemented to improve sleep or counteract alertness impairments in the shorter term.

Napping prior to shifts, particularly night shifts, is an important strategy for mitigating sleepiness, compensating for insufficient sleep and preventing extended wakefulness prior to shifts. In particular, the first night shift is associated with extended wakefulness (24, 54) when workers wake in the morning after the prior nocturnal sleep and remain awake during the day until the start of the first night shift. An afternoon/evening nap before the shift prevents workers from being awake for >24 hours at the end of the first night shift. This is a commonly adopted strategy, with almost 50% of shift workers reporting an afternoon/evening nap before the first night shift (55). Between consecutive night shifts, although workers are restricted to daytime sleep opportunities and the duration of wakefulness tends to be reduced, one-third of workers report napping prior to the subsequent night shifts (55).

Naps during night shifts can also improve alertness (56), with even short naps (<20 minutes) effective for improving alertness and reaction times (57). These can be particularly valuable for preventing extended wakefulness during long shifts (e.g., 12-hour shifts), which are common across industries. The benefits of napping, however, should be carefully balanced with the consequences related to sleep inertia (58), a common experience of grogginess upon awakening (59). Sleep inertia is particularly pronounced following sleep deprivation, and when waking during the biological night (60), where it can occur even after relatively short sleep episodes or naps (61). Combined with chronic sleep deprivation and circadian misalignment, sleep inertia can create the perfect recipe for workplace errors and accidents (53). Thus, where there are opportunities for naps during shifts, time should be allocated for the dissipation of sleep inertia after waking, particularly in safety critical occupations that often require rapid, important decisions immediately after waking (62).

Caffeine consumption is a commonly recommended strategy to improve alertness during night shifts, and can be particularly useful for mitigating alertness impairments towards the end of night shifts (63, 64). Low to moderate doses of caffeine improve cognitive performance (65), whereas higher doses of caffeine may be counterintuitive as it may induce side effects (e.g., anxiety, jitteriness) (65). This, however, depends on whether individuals are well-rested or sleep-deprived. For well-rested individuals, alertness improvements typically occur with doses of 30–300 mg (66, 67), whereas higher doses (up to 600 mg) may be necessary for sleep-deprived individuals (68, 69). For night shift workers, rather than a single, high dose of caffeine, low, repeated doses across the shift are recommended (70). Caffeine consumption in the latter part of night shifts should be carefully managed, as it can disrupt subsequent daytime sleep (71). That being said, individual sensitivity to caffeine varies (72), and the optimal dosage, frequency, and timing for caffeine are therefore different for every individual.

Similar to caffeine, light exposure can acutely improve night shift workers' alertness (53, 73). Blue-enriched white light (higher colour temperature) is particularly effective (73, 74) as the human circadian system is maximally sensitive to short wavelength blue light (~480 nm) (75–79). These alerting effects are partially linked to suppression of the sleep-promoting hormone melatonin (80, 81). However, other neural pathways are also involved, as alertness improvements occur even during the daytime when melatonin levels are low (82, 83).

Reducing circadian misalignment via strategic light-dark exposure

For individuals working multiple consecutive or permanent night shifts, extended use of strategies that offer symptomatic relief is not advisable. These strategies do not address the underlying mechanism of circadian misalignment, and can exacerbate chronic sleep loss (e.g., prolonged use of caffeine can impair daytime sleep), particularly for individuals with low shift work tolerance (84). Instead, strategies to promote circadian adaptation should be applied to most reliably manage alertness impairments (53). For circadian adaptation to night shifts, for example, the goal may be to shift the phase of the circadian pacemaker to delay the peak of sleepiness (i.e., CBTmin) out of the night shift and into the daytime sleep episode (85). Doing so should improve shift workers' sleep quantity and quality. When individuals are entrained (i.e., the internal circadian period is synchronised to the period of the external environment period), CBTmin typically occurs somewhere between the start of the nocturnal sleep and 2–3 hours before the end of nocturnal sleep (18, 86). Under these principles, delaying night shift workers' CBTmin into this relative timing for daytime sleep episodes should improve sleep quantity and quality.

Not only does artificial light exposure acutely improve alertness as mentioned above, it is a feasible method to facilitate phase shifts in shift-working contexts. When administered at the appropriate circadian timing, even relatively short light exposure can induce phase shifts, as the circadian system exhibits a non-linear, dose-dependent response to light (87, 88). Short, intermittent pulses of bright light can reliably produce circadian phase shifts (89–91) and can be used in shift-working settings where extended exposure to bright light is not feasible. Furthermore, light exposure does not have to be at particularly high intensities, as melatonin suppression (50) and phase shifts (92, 93) occur with light as low as 10–12 lux.

Although individuals can adapt to night shift work by phase delaying in response to light exposure during the biological night, complete circadian adaptation is rare, even in permanent night shift workers (94, 95). Light in the morning during the post-shift commute home coincides with the phase-advance portion of the light PRC, therefore counteracting this circadian phase delay (96). Strategies to avoid light during the phase-advance portion of the light PRC, including wearing light-blocking sunglasses during the commute home, ensuring a dark home environment by drawing curtains, and sleeping as soon as possible after the shift, are all common strategies to prevent the phase-advancing morning sunlight and support subsequent daytime sleep (53, 97). The required 'intensity' of lighting strategies necessary to phase shift depends on individual circadian timing. A landmark study tested a combination of light strategies with five consecutive simulated night shifts (97). For individuals with later circadian timing, circadian adaptation was achieved by preventing phase-advancing light on the post-shift commute home. On the other hand, for individuals with earlier circadian timing, additional pulses of intermittent bright light on night shifts were required to confer the greater magnitude of phase delay necessary for circadian adaptation (97). Given that the commute home after night shifts is linked to adverse driving events, implementing light avoidance strategies in the real world requires careful consideration of safety risks. In practice, aiming for partial realignment of circadian rhythms with daytime sleep (i.e., phase delaying CBTmin into the first half of the daytime sleep) is more desirable than complete realignment (i.e., moving CBTmin into the second half of daytime sleep). Not only do they benefit alertness and performance to similar extents (98), complete circadian adaptation (particularly to night shift) results in a greater degree of circadian misalignment on days off when shift workers return to daytime activities and night-time sleep (99, 100).

Future directions: balancing research and practicality

The array of health and safety consequences and the low retention of shift workers across multiple industries (101) highlight the need for effective interventions to manage shift work. Shift workers represent a unique population who live under long-term circadian misalignment for whom traditional sleep and well-being interventions that are largely based on nocturnal sleep are not applicable (102). For example, cognitive behavioural therapy for insomnia is efficacious for treating insomnia across many populations (103) but not for shift workers (104). Traditional sleep practices have limited applicability in this population. For instance, sleep hygiene practices (e.g., following a consistent bedtime) are not practical for shift workers. Further, without adaptation to shift workers, typical components such as sleep restriction may elevate safety risks in a population that is already chronically restricted (105, 106).

Implementing personalised interventions

With a global move towards personalised medicine for individualised healthcare, it is expected that interventions for circadian misalignment will continue to become more tailored. Individual strategies that consider work rosters and lifestyles (e.g., domestic duties) are crucial but lacking. However, differences in shift rotations, organisational requirements and individual circumstances among shift workers remain a barrier for delivering strategies at a population level. Preliminary evidence supports personalised behavioural strategies (e.g., promoting sleep hygiene) for improving shift workers' insomnia, depression and anxiety symptoms, and highlight promise for personalised shift work and sleep-wake management tools (107, 108).

Evidence on the predictors of individual differences in shift work response have limited translation into interventions. Sex difference in the response to shift work is one example. Females are more likely to have greater family commitments, primary caregiving roles and domestic duties which (1) further restrict time available to sleep and recover from shift work (109–111), (2) largely dictate sleep-wake timings (e.g., scheduling sleep around domestic duties), and (3) subsequently dictate light-exposure patterns. To address this, practical strategies to prioritise and protect time for sleep (e.g., strategic napping) (112) should be a key priority in industries with a high proportion of female workers (e.g., healthcare). On the other hand, male shift workers are more likely to report greater engagement in unhealthy lifestyle behaviours such as smoking and alcohol consumption when compared to their non-shift working counterparts (113). Interventions for managing health in male shift workers should therefore target such lifestyle behaviours.

One of the key challenges to enabling personalised interventions is the assessment of individual circadian phase. Reliable, real-time methods of measuring circadian rhythms in the field are limited. The gold-standard measurement relies on urinary samples over 24–48 hours to measure the rhythm of urinary 6-sulphatoxymelatonin, the metabolite of melatonin. Biomathematical models of sleep and alertness can be deployed to predict circadian phase, offering a potential solution (114, 115). These models, however, were largely developed and validated in populations of relatively young, healthy Caucasian adults with regular, diurnal sleep-wake schedules, requiring further research to examine their stability in predicting circadian phase in shift working populations. These models also traditionally consider photopic illuminance when predicting the circadian response to light (116). There is promise to improve individual phase predictions by incorporating light sensitivity (116) and utilising melanopic illuminance to more accurately quantify the amount of light stimulating the circadian photosensitive receptor melanopsin (75, 117).

Digitised delivery of sleep health interventions in shift workers

Adherence to face-to-face interventions in shift workers is low (107), and access to standard healthcare services is challenging due to shift workers' non-standard working hours. Telehealth and app-based services are increasingly used for managing sleep disturbances and transmeridian travel (118, 119). It is now well-established that the digitised delivery of health support increases program retention and is particularly beneficial for populations with limited time availability (120, 121), offering a promising avenue for delivering accessible, convenient support to shift workers.

Across multiple populations, prescription of sleep schedules has shown the greatest potential for improving sleep when compared to other methods such as coaching or sleep education (122). This can be of particular benefit for shift workers, who may be aware of the strategies, but face barriers to implementing them (e.g., optimal sleep-wake timings or caffeine dosage). App-based tools have the potential to combine multiple intervention modalities, including the use of biomathematical models to predict phase and provide personalised recommendations. These tools can also help with goal setting, incentivising and self-monitoring, which are known to produce positive changes in sleep in general populations (123). A recent pilot study has shown positive uptake of circadian health management tools in shift workers, demonstrating improvements in sleep-related impairments, insomnia symptoms and mental health (108). The effectiveness and long-term adoption of telehealth, web or app-based services for shift workers requires further investigation and rigorous testing. If successful, these modes can offer more scalable solutions for addressing health and circadian misalignment in shift workers and broader populations.

Summary

Shift workers often experience chronic circadian misalignment and associated sleep disruption. Traditional strategies for managing sleep and well-being that are designed for day-working, nocturnal sleepers, are rarely applicable or efficacious for this population. The substantial individual differences in shift work tolerance and circadian response to shift work, alongside the wide range of health and safety consequences, and high turnover rate across various industries, beg the need for effective, personalised shift work management strategies. Current strategies for managing shift work are typically based on a 'one-size-fits-all' approach and do not consider lifestyle factors, which reduces their applicability long-term. There is potential to leverage biomathematical modelling and digital technologies to deliver personalised interventions that are accessible, convenient, and provide timely support to shift workers.

References

1 Czeisler CA, Duffy JF, Shanahan TL, Brown EN, Mitchell JF, Rimmer DW, et al. Stability, precision, and near-24-hour period of the human circadian pacemaker. Science. 1999;284(5423):2177–81.
2 Cajochen C, Khalsa SB, Wyatt JK, Czeisler CA, Dijk DJ. EEG and ocular correlates of circadian melatonin phase and human performance decrements during sleep loss. Am J Physiol. 1999;277(3 Pt 2):R640–9.
3 Lewy A, Emens J, Songer J, Rough J. The neurohormone melatonin as a marker, medicament, and mediator. In: Hormones, Brain and Behavior Online. Elsevier Inc.; 2010. p. 2505–28. Available from: https://ohsu.pure.elsevier.com/en/publications/the-neurohormone-melatonin-as-a-marker-medicament-and-mediator-2

4 Mistlberger RE. Circadian regulation of sleep in mammals: role of the suprachias-matic nucleus. Brain Res Brain Res Rev. 2005;49(3):429–54.

5 Yamazaki S, Numano R, Abe M, Hida A, Takahashi R, Ueda M, et al. Reset-ting central and peripheral circadian oscillators in transgenic rats. Science. 2000;288(5466):682–5.

6 Yoo S-H, Yamazaki S, Lowrey PL, Shimomura K, Ko CH, Buhr ED, et al. PERIOD2: LUCIFERASE real-time reporting of circadian dynamics reveals per-sistent circadian oscillations in mouse peripheral tissues. Proc Natl Acad Sci U S A. 2004;101(15):5339–46.

7 Burgess HJ, Eastman CI. The dim light melatonin onset following fixed and free sleep schedules. J Sleep Res. 2005;14(3):229–37.

8 Zisapel N. New perspectives on the role of melatonin in human sleep, circadian rhythms and their regulation. Br J Pharmacol. 2018;175(16):3190–9.

9 Czeisler CA, Weitzman ED, Moore-Ede MC, Zimmerman JC, Knauer RS. Human sleep: its duration and organization depend on its circadian phase. Science. 1980;210(4475):1264–7.

10 Dijk DJ, Cajochen C. Melatonin and the circadian regulation of sleep initiation, consolidation, structure, and the sleep EEG. J Biol Rhythms. 1997;12(6):627–35.

11 Arendt J. Shift work: coping with the biological clock. Occup Med. 2010;60(1):10–20.

12 Sharma VK, Chandrashekaran MK. Zeitgebers (time cues) for biological clocks. Curr Sci. 2005;89(7):1136–46.

13 Brown EN, Czeisler CA. The statistical analysis of circadian phase and amplitude in constant-routine core-temperature data. J Biol Rhythms. 1992;7(3):177–202.

14 Czeisler CA, Gooley JJ. Sleep and circadian rhythms in humans. Cold Spring Harb Symp Quant Biol. 2007;72:579–97.

15 Stone JE, Sletten TL, Magee M. Temporal dynamics of circadian phase shifting response to consecutive night shifts in healthcare workers: role of light – dark expo-sure. J Physiol. 2018;596(12):2381–95.

16 Gooley JJ, Chamberlain K, Smith KA, Khalsa SBS, Rajaratnam SMW, Van Reen E, et al. Exposure to room light before bedtime suppresses melatonin onset and shortens melatonin duration in humans. J Clin Endocrinol Metab. 2011;96(3):E463–72.

17 Khalsa SBS, Jewett ME, Cajochen C, Czeisler CA. A phase response curve to single bright light pulses in human subjects. J Physiol. 2003;549(Pt 3):945–52.

18 Wever RA. The Circadian System of Man: Results of Experiments Under Tem-poral Isolation. Springer Science & Business Media; 2013. 276 p. Available from: https://play.google.com/store/books/details?id=CVfmBwAAQBAJ

19 Cho S-S, Lee D-W, Kang M-Y. The association between shift work and health-related productivity loss due to either sickness absence or reduced performance at work: a cross-sectional study of Korea. Int J Environ Res Public Health. 2020;17(22).

20 Andrzejczak D, Kapała-Kempa M, Zawilska JB. Health consequences of shift work. Przegl Lek. 2011;68(7):383–7.

21 Kecklund G, Axelsson J. Health consequences of shift work and insufficient sleep. BMJ. 2016;355:i5210.

22 Åkerstedt T. Shift work and disturbed sleep/wakefulness. Occup Med. 2003;53(2):89–94.

23 Akerstedt T, Wright KP Jr. Sleep loss and fatigue in shift work and shift work dis-order. Sleep Med Clin. 2009;4(2):257–71.

24 Ganesan S, Magee M, Stone JE, Mulhall MD, Collins A, Howard ME, et al. The impact of shift work on sleep, alertness and performance in healthcare workers. Sci Rep. 2019;9(1):4635.

25 Khan WAA, Conduit R, Kennedy GA, Jackson ML. The relationship between shift-work, sleep, and mental health among paramedics in Australia. Sleep Health. 2020;6(3):330–7.

26 Kang M-Y, Kwon H-J, Choi K-H, Kang C-W, Kim H. The relationship between shift work and mental health among electronics workers in South Korea: a cross-sectional study. PLoS ONE. 2017;12(11):e0188019.

27 Vyas MV, Garg AX, Iansavichus AV, Costella J, Donner A, Laugsand LE, et al. Shift work and vascular events: systematic review and meta-analysis. BMJ. 2012;345:e4800.

28 Wegrzyn LR, Tamimi RM, Rosner BA, Brown SB, Stevens RG, Eliassen AH, et al. Rotating night-shift work and the risk of breast cancer in the nurses' health studies. Am J Epidemiol. 2017;186(5):532–40.

29 Dembe AE, Erickson JB, Delbos RG, Banks SM. Nonstandard shift schedules and the risk of job-related injuries. Scand J Work Environ Health. 2006;32(3):232–40. http://doi.org/10.5271/sjweh.1004

30 Ftouni S, Sletten TL, Howard M, Anderson C, Lenné MG, Lockley SW, et al. Objective and subjective measures of sleepiness, and their associations with on-road driving events in shift workers. J Sleep Res. 2013;22(1):58–69.

31 Ftouni S, Sletten TL, Nicholas CL, Kennaway DJ, Lockley SW, Rajaratnam SMW. Ocular measures of sleepiness are increased in night shift workers undergoing a simulated night shift near the peak time of the 6-sulfatoxymelatonin rhythm. J Clin Sleep Med. 2015;11(10):1131–41.

32 Mulhall MD, Sletten TL, Magee M, Stone JE, Ganesan S, Collins A, et al. Sleepiness and driving events in shift workers: the impact of circadian and homeostatic factors. Sleep. 2019;42(6):zsz074.

33 Andlauer P, Reinberg A, Fourré L, Battle W, Duverneuil G. Amplitude of the oral temperature circadian rhythm and the tolerance to shift-work. J Physiol. 1979;75(5):507–12.

34 Degenfellner J, Schernhammer E. Shift work tolerance. Occup Med. 2021. http://doi.org/10.1093/occmed/kqab138

35 Flo E, Pallesen S, Magerøy N, Moen BE, Grønli J, Hilde Nordhus I, et al. Shift work disorder in nurses – assessment, prevalence and related health problems. PLoS ONE. 2012;7(4):e33981.

36 Waage S, Pallesen S, Moen BE, Magerøy N, Flo E, Di Milia L, et al. Predictors of shift work disorder among nurses: a longitudinal study. Sleep Med. 2014;15(12):1449–55.

37 American Psychiatric Association. Diagnostic and Statistical Manual of Mental Disorders (IV-TR). 4th ed. Washington, DC: American Psychiatric Association; 2000.

38 Martin JS, Laberge L, Sasseville A, Bérubé M, Alain S, Houle J, et al. Day and night shift schedules are associated with lower sleep quality in Evening-types. Chronobiol Int. 2015;32(5):627–36.

39 Archer SN, Robilliard DL, Skene DJ, Smits M, Williams A, Arendt J, et al. A length polymorphism in the circadian clock gene Per3 is linked to delayed sleep phase syndrome and extreme diurnal preference. Sleep. 2003;26(4):413–15.

40 Drake CL, Belcher R, Howard R, Roth T, Levin AM, Gumenyuk V. Length polymorphism in the Period 3 gene is associated with sleepiness and maladaptive circadian phase in night-shift workers. J Sleep Res. 2015;24(3):254–61.

41 Admi H, Tzischinsky O, Epstein R, Herer P, Lavie P. Shift work in nursing: is it really a risk factor for nurses' health and patients' safety? Nurs Econ. 2008;26(4):250–7.

42 Åhsberg E, Kecklund G, Åkerstedt T, Francesco Gamberale. Shiftwork and different dimensions of fatigue. Int J Ind Ergon. 2000;26(4):457–65.

43 Bara A-C, Arber S. Working shifts and mental health – findings from the British Household Panel Survey (1995–2005). Scand J Work Environ Health. 2009;35(5):361–7.

44 Rouch I, Wild P, Ansiau D, Marquié J-C. Shiftwork experience, age and cognitive performance. Ergonomics. 2005;48(10):1282–93.

45 Kecklund G, Eriksen CA, Akerstedt T. Police officers attitude to different shift systems: association with age, present shift schedule, health and sleep/wake complaints. Appl Ergon. 2008;39(5):565–71.

46 Seo Y-J, Matsumoto K, Park Y-M, Shinkoda H, Noh T-J. The relationship between sleep and shift system, age and chronotype in shift workers. Biol Rhythm Res. 2000;31(5):559–79.

47 Burch JB, Tom J, Zhai Y, Criswell L, Leo E, Ogoussan K. Shiftwork impacts and adaptation among health care workers. Occup Med. 2009;59(3):159–66.

48 Härmä M, Tarja H, Irja K, Mikael S, Jussi V, Anne B, et al. A controlled intervention study on the effects of a very rapidly forward rotating shift system on sleep – wakefulness and well-being among young and elderly shift workers. Int J Psychophysiol. 2006;59(1):70–9.

49 Dumont M, Benhaberou-Brun D, Paquet J. Profile of 24-h light exposure and circadian phase of melatonin secretion in night workers. J Biol Rhythms. 2001;16(5):502–11.

50 Phillips AJK, Vidafar P, Burns AC, McGlashan EM, Anderson C, Rajaratnam SMW, et al. High sensitivity and interindividual variability in the response of the human circadian system to evening light. Proc Natl Acad Sci U S A. 2019;116(24):12019–24.

51 Harris R, Beatty CJ, Cori JM, Spitz G, Soleimanloo SS, Peterson SA, et al. The impact of break duration, time of break onset, and prior shift duration on the amount of sleep between shifts in heavy vehicle drivers. J Sleep Res. 2022;e13730.

52 Roach GD, Reid KJ, Dawson D. The amount of sleep obtained by locomotive engineers: effects of break duration and time of break onset. Occup Environ Med. 2003;60(12):e17.

53 Smith MR, Eastman CI. Shift work: health, performance and safety problems, traditional countermeasures, and innovative management strategies to reduce circadian misalignment. Nat Sci Sleep. 2012;4:111–32.

54 Folkard S. Is there a 'best compromise' shift system? Ergonomics. 1992;35(12):1453–63; discussion 1465–6.

55 Axelsson J, Salllinen M, Sundelin T, Kecklund G. Sleep and shift work. In: Cappuccio FP, Miller MA, Lockley SW, Rajaratnam SMW, editors. Sleep, Health, and Society: From Aetiology to Public Health. 2nd ed. London: Oxford University Press; 2018. p. 179–88.

56 Bonnefond A, Muzet A, Winter-Dill AS, Bailloeuil C, Bitouze F, Bonneau A. Innovative working schedule: introducing one short nap during the night shift. Ergonomics. 2001;44(10):937–45.

57 Purnell MT, Feyer AM, Herbison GP. The impact of a nap opportunity during the night shift on the performance and alertness of 12-h shift workers. J Sleep Res. 2002;11(3):219–27.

58 Kubo T, Takahashi M, Takeyama H, Matsumoto S, Ebara T, Murata K, et al. How do the timing and length of a night-shift nap affect sleep inertia? Chronobiol Int. 2010;27(5):1031–44.

59 Silva EJ, Duffy JF. Sleep inertia varies with circadian phase and sleep stage in older adults. Behav Neurosci. 2008;122(4):928–35.

60 Scheer FAJL, Shea TJ, Hilton MF, Shea SA. An endogenous circadian rhythm in sleep inertia results in greatest cognitive impairment upon awakening during the biological night. J Biol Rhythms. 2008;23(4):353–61.

61 Hofer-Tinguely G, Achermann P, Landolt H-P, Regel SJ, Rétey JV, Dürr R, et al. Sleep inertia: performance changes after sleep, rest and active waking. Cogn Brain Res. 2005;22(3):323–31.

62 Ruggiero JS, Redeker NS. Effects of napping on sleepiness and sleep-related performance deficits in night-shift workers: a systematic review. Biol Res Nurs. 2014;16(2):134–42.

63 McLellan TM, Kamimori GH, Bell DG, Smith IF, Johnson D, Belenky G. Caffeine maintains vigilance and marksmanship in simulated urban operations with sleep deprivation. Aviat Space Environ Med. 2005;76(1):39–45.

64 McLellan TM, Kamimori GH, Voss DM, Bell DG, Cole KG, Johnson D. Caffeine maintains vigilance and improves run times during night operations for Special Forces. Aviat Space Environ Med. 2005;76(7):647–54.

65 McLellan TM, Caldwell JA, Lieberman HR. A review of caffeine's effects on cognitive, physical and occupational performance. Neurosci Biobehav Rev. 2016;71:294–312.

66 Smith AP. Caffeine at work. Hum Psychopharmacol. 2005;20(6):441–5.

67 Carvey C, Thompson L, Mahoney C, Lieberman H. Caffeine: mechanism of action, genetics, and behavioral studies conducted in task simulators and the field. In Sleep Deprivation, Stimulant Medications, and Cognition; 2012. p. 93–107.

68 Lieberman HR, Tharion WJ, Shukitt-Hale B, Speckman KL, Tulley R. Effects of caffeine, sleep loss, and stress on cognitive performance and mood during U.S. Navy SEAL training. Psychopharmacology. 2002;164(3):250–61.

69 Wesensten NJ, Belenky G, Kautz MA, Thorne DR, Reichardt RM, Balkin TJ. Maintaining alertness and performance during sleep deprivation: modafinil versus caffeine. Psychopharmacology. 2002;159(3):238–47.

70 Wyatt JK, Cajochen C, Ritz-De Cecco A, Czeisler CA, Dijk D-J. Low-dose repeated caffeine administration for circadian-phase-dependent performance degradation during extended wakefulness. Sleep. 2004;27(3):374–81.

71 Carrier J, Fernandez-Bolanos M, Robillard R, Dumont M, Paquet J, Selmaoui B, et al. Effects of caffeine are more marked on daytime recovery sleep than on nocturnal sleep. Neuropsychopharmacology. 2007;32(4):964–72.

72 Clark I, Landolt HP. Coffee, caffeine, and sleep: a systematic review of epidemiological studies and randomized controlled trials. Sleep Med Rev. 2017;31:70–8.

73 Wu C-J, Huang T-Y, Ou S-F, Shiea J-T, Lee B-O. Effects of lighting interventions to improve sleepiness in night-shift workers: a systematic review and meta-analysis. Healthcare (Basel). 2022;10(8):1390.

74 Sletten TL, Raman B, Magee M, Ferguson SA, Kennaway DJ, Grunstein RR, et al. A blue-enriched, increased intensity light intervention to improve alertness and performance in rotating night shift workers in an operational setting. Nat Sci Sleep. 2021;13:647–57.

75 Lucas RJ, Peirson SN, Berson DM, Brown TM, Cooper HM, Czeisler CA, et al. Measuring and using light in the melanopsin age. Trends Neurosci. 2014;37(1):1–9.

76 Brainard GC, Hanifin JP, Greeson JM, Byrne B, Glickman G, Gerner E, et al. Action spectrum for melatonin regulation in humans: evidence for a novel circadian photoreceptor. J Neurosci. 2001;21(16):6405–12.

77 Berson DM, Dunn FA, Takao M. Phototransduction by retinal ganglion cells that set the circadian clock. Science. 2002;295(5557):1070–3.

78 Gooley JJ, Lu J, Chou TC, Scammell TE, Saper CB. Melanopsin in cells of origin of the retinohypothalamic tract. Nat Neurosci. 2001;4(12):1165.

79 Panda S, Provencio I, Tu DC, Pires SS, Rollag MD, Castrucci AM, et al. Melanopsin is required for non-image-forming photic responses in blind mice. Science. 2003;301(5632):525–7.

80 Lewy AJ, Wehr TA, Goodwin FK, Newsome DA, Markey SP. Light suppresses melatonin secretion in humans. Science. 1980;210(4475):1267–9.

81 Pallesen S, Bjorvatn B, Magerøy N, Saksvik IB, Waage S, Moen BE. Measures to counteract the negative effects of night work. Scand J Work Environ Health. 2010;36(2):109–20.

82 Phipps-Nelson J, Redman JR, Dijk D-J, Rajaratnam SMW. Daytime exposure to bright light, as compared to dim light, decreases sleepiness and improves psychomotor vigilance performance. Sleep. 2003;26(6):695–700.

83 Vandewalle G, Balteau E, Phillips C, Degueldre C, Moreau V, Sterpenich V, et al. Daytime light exposure dynamically enhances brain responses. Curr Biol. 2006;16(16):1616–21.

84 Smith MR, Fogg LF, Eastman CI. Practical interventions to promote circadian adaptation to permanent night shift work: study 4. J Biol Rhythms. 2009;24(2):161–72.

85 Eastman CI, Martin SK. How to use light and dark to produce circadian adaptation to night shift work. Ann Med. 1999;31(2):87–98.

86 Zulley J, Wever R, Aschoff J. The dependence of onset and duration of sleep on the circadian rhythm of rectal temperature. Pflügers Archiv. 1981;391(4):314–18.

87 Boivin DB, Duffy JF, Kronauer RE, Czeisler CA. Dose-response relationships for resetting of human circadian clock by light. Nature. 1996;379(6565):540–2.

88 Zeitzer JM, Dijk DJ, Kronauer R, Brown E, Czeisler C. Sensitivity of the human circadian pacemaker to nocturnal light: melatonin phase resetting and suppression. J Physiol. 2000;526(Pt 3):695–702.

89 Chang A-M, Santhi N, St Hilaire M, Gronfier C, Bradstreet DS, Duffy JF, et al. Human responses to bright light of different durations. J Physiol. 2012;590(13):3103–12.

90 Rimmer DW, Boivin DB, Shanahan TL, Kronauer RE, Duffy JF, Czeisler CA. Dynamic resetting of the human circadian pacemaker by intermittent bright light. Am J Physiol Regul Integr Comp Physiol. 2000;279(5):R1574–9.

91 Gronfier C, Wright KP Jr, Kronauer RE, Jewett ME, Czeisler CA. Efficacy of a single sequence of intermittent bright light pulses for delaying circadian phase in humans. Am J Physiol Endocrinol Metab. 2004;287(1):E174–81.

92 Duffy JF, Wright KP Jr. Entrainment of the human circadian system by light. J Biol Rhythms. 2005;20(4):326–38.

93 Wright KP Jr, Czeisler CA. Absence of circadian phase resetting in response to bright light behind the knees. Science. 2002;297(5581):571.

94 Ferguson SA, Kennaway DJ, Baker A, Lamond N, Dawson D. Sleep and circadian rhythms in mining operators: limited evidence of adaptation to night shifts. Appl Ergon. 2012;43(4):695–701.

95 Folkard S. Do permanent night workers show circadian adjustment? A review based on the endogenous melatonin rhythm. Chronobiol Int. 2008;25(2):215–24.

96 Lunn RM, Blask DE, Coogan AN, Figueiro MG, Gorman MR, Hall JE, et al. Health consequences of electric lighting practices in the modern world: a report on the National Toxicology Program's workshop on shift work at night, artificial light at night, and circadian disruption. Sci Total Environ. 2017;607–8:1073–84.

97 Crowley SJ, Lee C, Tseng CY, Fogg LF, Eastman CI. Combinations of bright light, scheduled dark, sunglasses, and melatonin to facilitate circadian entrainment to night shift work. J Biol Rhythms. 2003;18(6):513–23.

98 Crowley SJ, Lee C, Tseng CY, Fogg LF, Eastman CI. Complete or partial circadian re-entrainment improves performance, alertness, and mood during night-shift work. Sleep. 2004;27(6):1077–87.

99 Lee C, Smith MR, Eastman CI. A compromise phase position for permanent night shift workers: circadian phase after two night shifts with scheduled sleep and light/dark exposure. Chronobiol Int. 2006;23(4):859–75.

100 Smith MR, Eastman CI. Night shift performance is improved by a compromise circadian phase position: study 3. Circadian phase after 7 night shifts with an intervening weekend off. Sleep. 2008;31(12):1639–45.

101 Sawatzky J-AV, Enns CL. Exploring the key predictors of retention in emergency nurses. J Nurs Manag. 2012;20(5):696–707.

102 Shriane AE, Ferguson SA, Jay SM, Vincent GE. Sleep hygiene in shift workers: a systematic literature review. Sleep Med Rev. 2020;53:101336.

103 Muench A, Vargas I, Grandner MA, Ellis JG, Posner D, Bastien CH, et al. We know CBT-I works, now what? Fac Rev. 2022;11:4.

104 Reynolds AC, Sweetman A, Crowther ME, Paterson JL, Scott H, Lechat B, et al. Is cognitive behavioral therapy for insomnia (CBTi) efficacious for treating insomnia symptoms in shift workers? A systematic review and meta-analysis. Sleep Med Rev. 2023;67:101716.

105 Järnefelt H, Lagerstedt R, Kajaste S, Sallinen M, Savolainen A, Hublin C. Cognitive behavioral therapy for shift workers with chronic insomnia. Sleep Med. 2012;13(10):1238–46.

106 Järnefelt H, Spiegelhalder K. CBT-I protocols for shift workers and health operators. In: Baglioni C, Espie CA, Riemann D, editors. Cognitive-Behavioural Therapy For Insomnia (CBT-I) Across The Life Span. Oxford: Wiley; 2022. p. 126–32.

107 Booker LA, Sletten TL, Barnes M, Alvaro P, Collins A, Chai-Coetzer CL, et al. The effectiveness of an individualized sleep and shift work education and coaching program to manage shift work disorder in nurses: a randomized controlled trial. J Clin Sleep Med. 2022;18(4):1035–45.

108 Murray JM, Magee M, Giliberto ES, Booker LA, Tucker AJ, Galaska B, et al. Mobile app for personalized sleep-wake management for shift workers: a user testing trial. Digit Health. 2023;9:20552076231165972.

109 Bergman B, Ahmad F, Stewart DE. Work family balance, stress, and salivary cortisol in men and women academic physicians. Int J Behav Med. 2008;15(1):54–61.

110 Clissold G, Smith P, Acutt B. The impact of unwaged domestic work on the duration and timing of sleep of female nurses working full-time on rotating 3-shift rosters. J Hum Ergol. 2001;30(1–2):345–9.

111 Spelten E, Totterdell P, Barton J, Folkard S. Effects of age and domestic commitment on the sleep and alertness of female shiftworkers. Work Stress. 1995;9(2–3):165–75.

112 Windred DP, Stone JE, McGlashan E, Cain SW, Phillips A. Attitudes towards sleep as a time commitment are associated with sleep regularity. Behav Sleep Med. 2021:1–12.

113 Ogeil RP, Savic M, Ferguson N, Lubman DI. Shift-work-play: understanding the positive and negative experiences of male and female shift workers to inform opportunities for intervention to improve health and wellbeing. Aust J Adv Nurs. 2021;38(2).

114 Knock SA, Magee M, Stone JE, Ganesan S, Mulhall MD, Lockley SW, et al. Prediction of shiftworker alertness, sleep, and circadian phase using a model of arousal dynamics constrained by shift schedules and light exposure. Sleep. 2021;44(11):zsab146

115 Stone JE, Postnova S, Sletten TL, Rajaratnam SMW, Phillips AJK. Computational approaches for individual circadian phase prediction in field settings. Curr Opin Syst Biol. 2020;22:39–51.

116 Stone JE, McGlashan EM, Quin N, Skinner K, Stephenson JJ, Cain SW, et al. The role of light sensitivity and intrinsic circadian period in predicting individual circadian timing. J Biol Rhythms. 2020;35(6):628–40.

117 Brown TM. Melanopic illuminance defines the magnitude of human circadian light responses under a wide range of conditions. J Pineal Res. 2020;69(1):e12655.

118 Rigney G, Walters A, Bin YS, Crome E, Vincent GE. Jet-lag countermeasures used by international business travelers. Aerosp Med Hum Perform. 2021;92(10):825–30.

119 Mason EC, Grierson AB, Sie A, Sharrock MJ, Li I, Chen AZ, et al. Co-occurring insomnia and anxiety: a randomized controlled trial of internet cognitive behavioral therapy for insomnia versus internet cognitive behavioral therapy for anxiety. Sleep. 2023;46(2):zsac205

120 Schueller SM, Torous J. Scaling evidence-based treatments through digital mental health. Am Psychol. 2020;75(8):1093–104.
121 Snoswell CL, Smith AC, Page M, Scuffham P, Caffery LJ. Quantifying the societal benefits from telehealth: productivity and reduced travel. Value Health Reg Issues. 2022;28:61–6.
122 Glazer Baron K, Culnan E, Duffecy J, Berendson M, Cheung Mason I, Lattie E, et al. How are consumer sleep technology data being used to deliver behavioral sleep medicine interventions? A systematic review. Behav Sleep Med. 2022;20(2):173–87.
123 Ong JL, Massar SAA, Lau TY, Ng BKL, Chan LF, Koek D, et al. A Randomised controlled trial of a digital, small incentive-based intervention for working adults with short sleep. Sleep. 2023;46(5):zsac315.

11

HEALTH DISPARITIES IN SLEEP AND MENTAL HEALTH

Examining the role of sleep disturbances in the relationship between climate change-related traumatic childhood experiences and mental health as an exemplar

Symielle A. Gaston, Rupsha Singh and Chandra L. Jackson

Introduction

Insufficient sleep is a global public health problem (1). For example, in 2006, the Institute of Medicine declared sleep deprivation and disorders as an unmet public health problem in the United States (US) (2). Despite goals to address poor sleep health, it has been estimated that only two-thirds of US adults obtain the recommended sleep duration, while even fewer children and adolescents (e.g., three out of ten high school-aged) obtain the recommended amount of sleep considered optimal for health and well-being (3, 4). Additionally, it has been consistently reported that certain populations are disproportionately burdened by poor sleep, including the socioeconomically disadvantaged, minoritised racial/ethnic groups and women (5, 6). Studies published since the COVID-19 pandemic also indicate that poor sleep health and sleep health disparities have been exacerbated (7), further highlighting that poor sleep is an urgent public health problem.

Similar to sleep health, mental health has likely worsened in recent years. Mental health disorders, including depressive, anxiety and substance use disorders, were three of the top five causes of years lived with disability in the Americas, even prior to the pandemic (8). Since the COVID-19 pandemic, one systematic review suggested the pooled prevalence of mental health issues was higher than the pre-pandemic prevalence with an estimated global prevalence of 28% for depression, 27% for anxiety and 28% for sleep problems (9).

DOI: 10.4324/9781003296966-11

Another systematic review and meta-analysis reported an even higher prevalence of sleep disturbances of 41% globally and suggested children along with adolescents were among the most affected groups during the pandemic (10). Examination of trend and cohort survey data in the US indicates the prevalence of depression and anxiety increased by approximately 30%–50% since the start of COVID-19 (at similar proportions among men and women), increased more among younger compared to older individuals (<60 versus 60+ years), and increased more among non-Hispanic (NH)-Black and Hispanic/Latino than NH-White people (11). To better elucidate mental health disparities, investigations across groups with multiple marginalised social identities (e.g., low-income Asian women) are warranted. Concurrent with COVID-19, other contemporary stressors such as war conflict, racial tensions and climate change across the globe may also adversely impact both sleep and mental health, as well as health disparities, in an additive or synergistic manner.

Given the impact and increasing prevalence of mental health disorders across the world, particularly among children and adolescents, as well as the potential exacerbation of disparities since the COVID-19 pandemic, it is important to understand and intervene upon the modifiable factors that reduce the burden of poor mental health. Sleep health, which is sensitive to early life adversity (12, 13), is one such modifiable factor that should be prioritised since it is increasingly recognised as an essential pillar of mental and physical health as well as a contributor to health disparities. Sleep health is also highly comorbid – in a bidirectional manner – with mental health outcomes (4, 14).

In this chapter, we illuminate the role of sleep disturbances in the relationship between traumatic events during childhood and mental health disparities by socioeconomic status, race/ethnicity and gender. We focus on climate change – an existential and emerging public health threat – as a traumatic childhood experience given its potential impact on sleep and mental health, especially among children (15–17). We also focus on depression and anxiety as mental health outcomes. Therefore, the role of sleep health in relationships between climate change-related childhood trauma and both depression and anxiety during adulthood is elucidated as a health disparities exemplar.

Sleep health disparities

A health disparity has been defined as 'a health difference that adversely affects defined disadvantaged populations, based on one or more health outcomes' (18). Outcomes identified in this definition include (1) higher incidence or prevalence, including earlier onset or more aggressive progression; (2) premature or excess mortality from specific conditions; (3) greater global burden of disease; (4) poorer health behaviours and clinical outcomes; (5) worse outcomes on validated self-reported measures that reflect daily functioning or symptoms from specific conditions (18). The US National Institute on Minority Health and Health Disparities considers the following as populations experiencing health disparities in the US: minoritised racial/ethnic groups (i.e., Asian American, Black/African American, Native American/Alaska Native, Native Hawaiian/Pacific Islander); underserved rural residents; individuals with low socioeconomic status; minoritised sexual and gender groups; and other populations that experience discrimination and social disadvantage (18). Marginalised or historically excluded groups in other countries can be considered populations disproportionately affected by health disparities. To overcome health disparities, equitable access to resources and opportunities for all people to live their healthiest lives is paramount (19).

According to health disparities definitions and criteria, sleep disparities exist in both paediatric and adult populations (5, 6, 20–26). Although more longitudinal studies using objective measures are needed, socioeconomic disparities in sleep have been observed among children, adolescents, and adults (26–28). Specifically, socioeconomic status (SES) indices including educational attainment, income, wealth/assets, occupation and employment status as well as neighbourhood SES factors have been associated with multiple sleep dimensions. In a review of studies using objective sleep measures, lower SES was associated with lower total sleep time (TST) and longer duration of wake after sleep onset (26). Higher SES was associated with shorter sleep latency, lower variability in sleep onset and higher sleep efficiency. Racial/ethnic disparities in sleep have also been consistently observed across the life course (5, 21–23, 29–31). Studies – mostly cross-sectional with self-reported data – were most conclusive about disparities in short sleep among NH-Black and Latinx individuals compared to their NH-White counterparts with less evidence (and fewer studies) for other minoritised racial/ethnic groups (e.g., Asian) (5, 21–23, 30, 31). While fewer studies have focused on sleep quality (e.g., lower sleep efficiency and poorer sleep architecture) and disorders, most found that minoritised racial/ethnic groups had the worst sleep quality, especially NH-Black adults who were less likely to report sleep complaints but had overall poorer objective sleep quality compared to NH-White adults (5, 21–23, 30, 31). Several gender disparities in sleep have also been observed. For instance, men have higher rates of sleep apnoea and women are more likely to report insomnia along with night-time sleep disturbances including short sleep (6, 24). Beyond social factors related to gender (e.g., caregiving) that can differentially impact sleep, reproductive hormones are hypothesised to contribute to sleep difference by sex/gender (24). Sex and reproductive hormones influence each of the two biological systems that stimulate sleep, the circadian system and the homeostatic drive for sleep, and are hypothesised to partially explain sex differences in the sleep-wake cycle (24).

Although we focus on socioeconomic, racial/ethnic and gender disparities for which the literature is most well-established, there are also observed disparities among other groups (25). For instance, most of the available literature among adolescents and adults (nine of ten studies in a recent review) generally report poorer sleep health among lesbian, gay and bisexual people than their heterosexual counterparts (25), which is of public health importance, but beyond the scope of this chapter.

Physical and mental health consequences of poor sleep and sleep disparities

Consequences of poor sleep on health

Poor sleep health has consequences for overall physical (for more details, see Chapter 8) and mental health along with well-being over the life course (2, 4, 32–35). Specifically, insufficient sleep duration, irregular sleep timing, poor sleep quality and sleep/circadian disorders are associated with adverse immunologic, metabolic, cardiovascular and neurocognitive health, which can contribute to the exacerbation of existing physical and mental health disparities across groups at varying risk of poor sleep (2, 4, 32–37). Further, intervention studies suggest causal relationships between sleep quality and mental health, such that improving sleep results in improvements of mood and anxiety symptoms (34, 35).

Consequences of disparities in poor sleep on physical health disparities

Marginalised groups often manifest disease at earlier life stages, and health conditions are often more severe with more aggressive progression (18, 26). Physical health disparities are likely explained, in part, by sleep disparities. For instance, poor sleep health may fundamentally contribute to cardiovascular health disparities since insufficient sleep, sleep disturbances and sleep disorders are associated with indicators of and risk factors for cardiovascular health (e.g., obesity, hypertension, type II diabetes), and both poor sleep and cardiovascular disease are widely observed to disproportionately impact marginalised groups (30, 38).

Consequences of disparities in poor sleep on mental health disparities

Associations between sleep health disparities and mental health disparities are not well-established. Studies around the world have consistently shown that lower income and education are major barriers to accessing mental health services (39) but, to our knowledge, have not explored sleep disparities as contributors to mental health disparities by SES. Further, the few identified studies outside the US were inconsistent about the existence of racial/ethnic disparities in mental health prevalence and services (40, 41). In the US where more studies have been conducted, data suggest similar or lower prevalence of psychiatric disorders among minoritised racial/ethnic groups compared to NH-Whites (42, 43); however, there are data limitations. First, symptom presentation may differ by race/ethnicity, with significantly higher somatic symptoms of depression among NH-Black and Latina compared to NH-White women as well as more functional impairments due to anxiety and depressive disorders along with more symptoms of distress among NH-Black compared to NH-White women (42–44), which may collectively preclude standard instruments from capturing overall depression. Relatedly, depression symptomology among minoritised racial/ethnic groups may include symptoms related to helplessness that are not as obviously linked to depression such as irritability, frustration, aggression, and violence (45). Therefore, variation in the manifestations of depression across racial/ethnic groups likely bias prevalence estimates (43). Second, similar prevalence by race/ethnicity may be an artefact of endorsement of psychiatric symptoms at higher thresholds among minoritised racial/ethnic groups (42, 43). Thirdly, cross-sectional prevalence measurements fail to capture potential differences in mental health status over time, thereby potentially differentially missing early-stage mental health disorders on the continuum to manifestation across social identity groups (42).

Despite the data inconsistencies regarding racial/ethnic disparities in mental health, mental health disparities are likely and can manifest in other forms. Racially/ethnically minoritised persons experience more risk factors (e.g., low SES) for poor mental health compared to non-Hispanic Whites (42) and face unique contributors (e.g., discrimination) to anxiety/depressive disorders (46), thus increasing the likelihood of these poor mental health outcomes within this group. Recent data shows that minoritised racial/ethnic groups have all experienced an increase in suicide concurrent with a decline in suicide among NH-White counterparts from 2018 to 2021 in the US (47). Such increases are likely related to the consistently observed identification and diagnosis, referral and treatment disparities in mental health as has been previously reported among racially/ethnically minoritised groups (42, 45, 48).

Unlike racial/ethnic disparities in mental health, studies more consistently conclude that women have twice the risk of anxiety and depression and higher lifetime prevalence of affective disorders compared to men (49–51). Further, symptom profiles of major depressive disorder differ between women and men. Specifically, women with major depressive disorder are more likely than their male counterparts to report experiences of sleep disturbances and have comorbid anxiety disorder (49, 52). Similar to hypothesised mechanisms explaining gender differences in sleep, researchers hypothesise sex hormone fluctuation as a major biological factor that may prime epigenetic influences on psychopathology and perpetuate sex as well as gender (e.g., menstruating individuals regardless of gender identity) differences in depression and anxiety (49, 51). In the social context, a meta-analysis also supported gender roles as a contributor to gender disparities in depression and concluded that androgyny (non-differentiated gender role) is protective against depression in both men and women (53). Importantly, individuals at the intersection of gender identity and race/ethnicity (e.g., transgender and racially/ethnically minoritised) may be even more disproportionately affected by depression and anxiety.

Although few studies have connected sleep disparities to disparities in mood disorders, a meta-analysis of polysomnography characteristics of mood disorders found sleep alterations such as sleep continuity disturbances, sleep depth and alterations to rapid eye movement sleep in most mental health disorders including affective disorders (e.g., depression) and anxiety (14). Given the higher prevalence of poor sleep among individuals with low SES, minoritised racial/ethnic groups and women, it is plausible that sleep may also propel disparities in mood disorders, namely depression and anxiety.

Determinants of sleep disparities

Determinants of sleep health disparities are complex, multilevel, multifactorial and share the same causal pathways observed for health outcomes that have well-known disparities (6, 22). As described by the World Health Organization, social determinants of health (including sleep health) are defined as the non-medical factors that influence health outcomes, inclusive of the conditions in which people are born, develop, work, live and age and the broader set of forces and systems shaping such conditions (54). Categories of social determinants of health include education access and quality, economic stability, healthcare access and quality, neighbourhood and built environment, and the social and community context (54), which interact with biological, lifestyle and behavioural factors (18, 20, 55, 56). Structural racism, described as the interconnected, reinforcing institutions that include the totality of ways societies foster racial discrimination and inequity across systems and society (57), is considered the most pervasive and main source of health disparities in the US (57, 58). Other societal structures including structural sexism and income inequality also contribute to the inequitable distributions of the social determinants of health as well as environmental injustice (58).

Due to interactions between deleterious societal structures, social determinants of health differ by SES, race/ethnicity and gender, and these factors contribute to disparities in sleep health. For instance, structural racism contributes to inequitable policies, racial residential segregation of neighbourhood environments, poorer access to quality education, less economic stability and concentrated poverty, trauma, and lower access to quality healthcare, which are observed more often among minoritised individuals (57). These factors,

among other manifestations of structural racism, impact sleep health disparities (20, 55, 56). For example, low SES and minoritised racial/ethnic groups
are more likely to work in occupations requiring shift work, and shift work has
been associated with long sleep latency, decreased slow wave sleep (SWS), and
chronic sleep loss among other sleep outcomes (59). Further, structural racism, structural sexism and income inequality contribute to differences in the
neighbourhood and built environment as well as the community and social
context, which has implications for sleep. For instance, neighbourhood disadvantage and disorder as well as low neighbourhood social cohesion and lower
neighbourhood safety, which are more common among lower SES, racially/
ethnically minoritised, and women (e.g., single mothers) have been associated
with poorer sleep quality and short sleep duration, likely through psychological distress related to living in such environments (55, 56). Further, within
neighbourhood and home environments, physical environmental determinants that disproportionately affect low SES and the racially/ethnically minoritised also negatively impact sleep (27, 55, 56). These factors include exposure
to artificial light, extreme temperatures, noise pollution, poor air quality, lower
greenspace access, inhospitable neighbourhood-built environments, and limited access to recreational facilities and healthy food stores among other neighbourhood features, all of which have been associated with disrupted sleep (55,
56). Furthermore, sex/gender was identified as a moderator of relationships
between environmental factors (e.g., ambient noise and room temperature)
and PSG-assessed sleep (27). The burden of exposure to these sleep disrup-
tors in the physical environment is expected to increase because of climate
change, which may further exacerbate sleep health disparities disproportionately among vulnerable groups including children (15).

Climate-related disasters as traumatic childhood experiences (TCEs) and the role of sleep in the TCE-mental health relationship

As previously stated, adverse social and physical environments contribute to
poor sleep health and sleep health disparities (55, 56). Further, both sleep disparities and mood disorders share common risk factors including traumas that
disproportionately impact minoritised groups (45). Relatedly, both racially/
ethnically minoritised individuals and women have higher burdens of childhood trauma (13, 60). Trauma or stressful life events (along with genetic
factors and other individual-level factors) are the strongest predictors of major
depressive disorders (51, 61). One study suggests that 35% of the differences
in depression between men and women are due to higher rates of childhood
trauma in the form of sexual abuse among women (62). Moreover, the burden of trauma is not equitable across racial/ethnic groups of women as early
life exposures including trauma are hypothesised to contribute to sleep health
disparities among racially/ethnically minoritised women (21), which may also

contribute to mental health inequities. Rather than provide an exhaustive review of childhood trauma and an exhaustive review of the scientific literature of disparities in childhood trauma, sleep and mood disorders, we provide an overview of the increasing threat of climate change events as a salient example. Therefore, we focus on – as a health disparities exemplar – relationships between childhood trauma in the form of climate change events in relation depression and anxiety while identifying the potential mediating role of sleep.

Climate-related disaster as a traumatic childhood experience

Childhood trauma includes traditionally studied adverse childhood experiences (ACEs), at household and community levels, as well as other increasingly recognised but currently understudied traumatic childhood experiences (TCEs). Seminal work by Felitti et al. first introduced the concept of ACEs as life course risk factors for morbidity and mortality during adulthood: Seven categories of ACEs included psychological, physical or sexual abuse; violence against mother; or living in a household with members who abused substances, were mentally ill or suicidal, or were ever imprisoned. We refer to TCEs rather than ACEs to denote the expanded definition of childhood trauma, where applicable (63). Furthermore, Felitti et al. found associations between these ACEs and risk factors for mortality in adults, including depression (63). Additionally, in a recent systematic review of 23 articles published between 1990 and 2018, authors reported that ACEs contributed to approximately 30% of anxiety and 40% of depression cases (64).

Prior literature has established ACEs as risk factors for poor mental health during adulthood; yet, children also face other TCEs worthy of investigation and intervention, namely climate-related disaster (16). Climate change – caused by greenhouse gas emissions due to human activity – affects global temperature and precipitation. Changes to temperature and precipitation contribute to geophysical, hydrologic, meteorologic and climatologic natural disasters and severe weather events (e.g., droughts, heat waves, storms floods, – see Hidalgo et al. for a complete list), their frequency, duration, and severity (65, 66). Such events have direct and indirect impacts on the physical (e.g., heat stress) and social environments (e.g., displacement/forced migration post-disaster) (65, 67) as well as psychosocial impacts such as community disruption (e.g., civil conflicts) (68). Other direct and indirect consequences of climate events also include, for instance, air pollution, water-borne and vector-borne disease, food insecurity, negative economic impacts, etc. (see Figure 11.1) (69, 70). Even in the absence of experiencing actual disaster, climate change can contribute to emotional distress related to awareness of climate change. For example, media messages may heighten stress responses to climate change (71). Relatively newly created terms for such distress include 'psychoterratic syndromes', which include 'ecoanxiety' – the anxiety people face upon

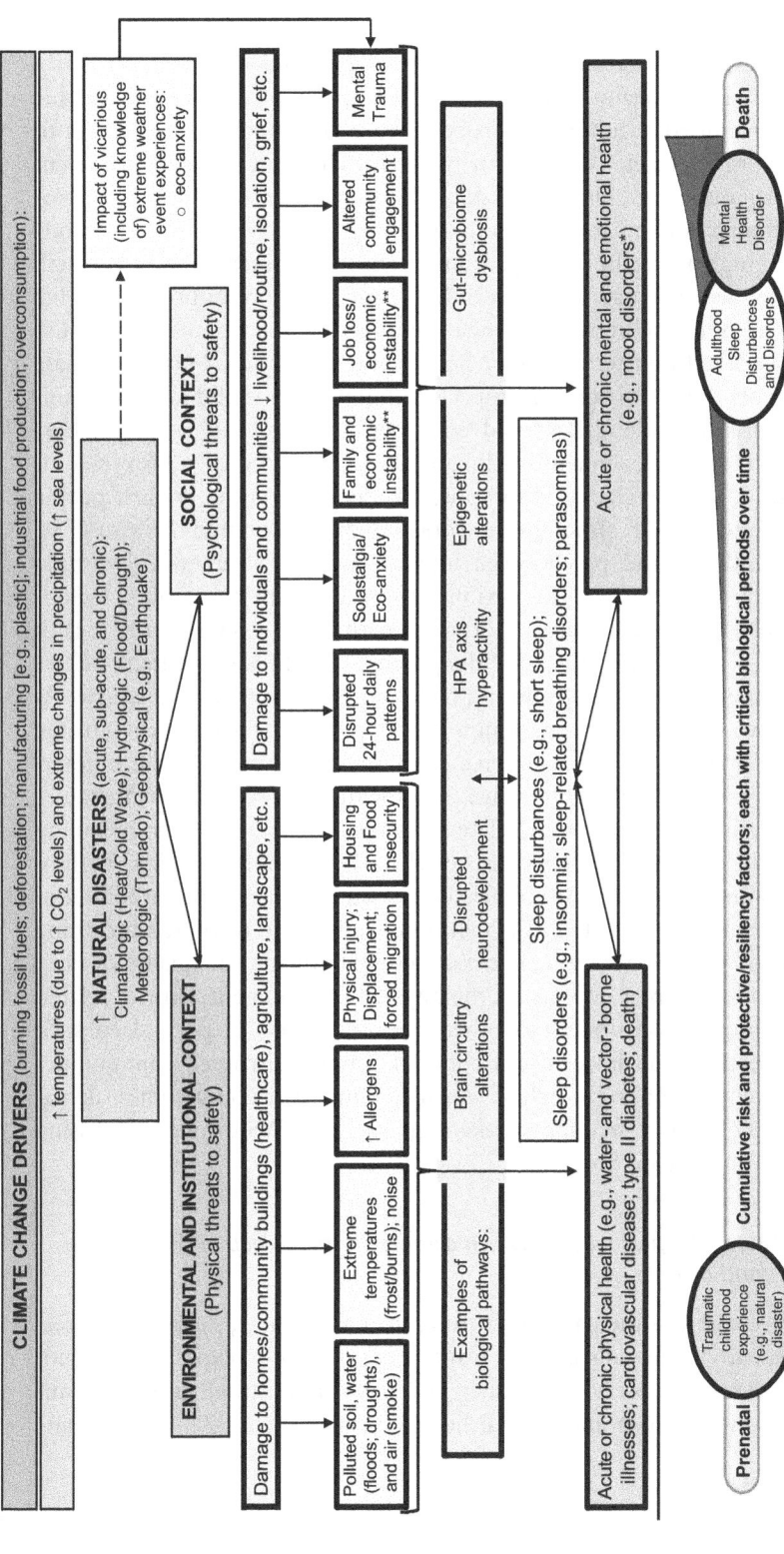

Figure 1. Conceptual Framework of Pathways from Climate Change Drivers to Poor Sleep and Poor Physical and Mental Health across the Life Course

*Major depressive disorder, generalized anxiety disorder, post-traumatic stress disorder, etc.; ** Economic instability could lead to community violence/civil conflict/war

FIGURE 11.1 Conceptual Framework of Pathways from Climate Change Drivers to Poor Sleep and Poor Physical and Mental Health across the Life Course.

being engulfed in threats related to climate change and 'solastalgia' – psychological distress from degradation of one's home environment (69, 70).

Children are especially vulnerable to the impacts of climate change due to their (1) limited autonomy and dependence on caregivers and adults to set and enforce policies to protect them against climate change; (2) different health behaviours (e.g., playing outside) that may increase exposure; (3) biological differences compared to adults (e.g., higher exposures per unit body weight and higher metabolic rate, advancement through rapid rates of growth and development during exposure windows) and (4) more future years of life compared to adults allowing for longer periods of latency between exposure to climate-related disaster and poor health outcomes (72). Further, climate change events disproportionately impact both low-income and racially/ethnically minoritised children due to adverse conditions and higher disease burdens prior to and independent of climate change (60, 65, 67). Also relevant to disparities, climate change disproportionately impacts the poorest people first, worst and longest, although the poorest have contributed the least (73).

Direct, indirect and psychosocial impacts as well as emotional distress related to climate change can contribute to poor health behaviours, including sleep (15), and have been reported as pathways to poor mental health (65, 67–70, 74–76) and, thus, have the potential do disrupt children's behavioural and mental health both during childhood and over the life course. As an illustrative example, extreme temperatures resultant from climate change can negatively impact sleep maintenance or the ability to maintain uninterrupted sleep, which is highly sensitive to ambient and core body temperature (15, 55). Other climate change-related environmental pollutants, including artificial light at night, not only disrupt sleep by altering circadian rhythms and melatonin production (55) but also contribute to increasing carbon monoxide in the atmosphere (15). Air pollution is related to disrupted breathing during sleep and increased wakefulness, thereby reducing the quality of sleep (55). Beyond the direct impacts on the physical environment, climate-related disaster indirectly contributes to other stressors linked to poor sleep such as food insecurity and lower social cohesion among disrupted communities post-displacement (65, 67), each of which can impact sleep outcomes such as poorer sleep efficiency, frequent awakenings and less SWS through increasing psychosocial stress (55, 56).

Potential biological pathways between climate change, sleep and depression and anxiety

Although research of the pathway between climate change, sleep and depression along with anxiety disorders is scant, it is plausible to expect similar biological mechanisms observed for other TCEs and sleep as well as mechanisms observed between sleep and mental health outcomes (65, 77–80). Climate change has environmental and social consequences (e.g., air, water, and soil pollution; economic instability), which have direct impacts on biological systems, thereby impacting sleep as well as physical and mental health over the life course (see Figure 11.1) (65, 77). Climate change contributes to natural disasters, which have direct and indirect effects on the physical and social environment as well as distress related to vicarious climate change experiences. Such adversity during early childhood has been associated with activation of stress response pathways, including both hyper- and hypoactivity of the hypothalamic–pituitary–adrenal (HPA) axis, which are marked by elevated and suppressed levels of corticotropin-releasing hormone (CRH) and cortisol (78, 79), respectively. Dysregulation of circadian rhythms of cortisol secretion (e.g., elevated cortisol levels at night) can result in sleep disruptions and alterations in allostasis (13, 81). Further, prolonged HPA axis activation can contribute to excess glucocorticoid secretion and oxidative stress, which is hypothesised to accelerate biological ageing (80). Although limited, emerging research demonstrates that epigenetic alterations, specifically DNA methylation, are associated with TCEs, sleep disturbance and psychopathology (82–85). Other (non-exhaustive) potential pathways illustrated in Figure 11.1 include the impacts of TCEs on brain development, including deleterious effects on structural, neuronal and functional development of the brain (82, 83, 86). There is increasing evidence that the gut–brain axis may serve as an important pathway between TCEs, sleep and mood disorders. In particular, gut microbiome dysbiosis, influenced by environmental and lifestyle factors, appear to be associated with both sleep disorders and mental health disorders such as depression (87, 88).

Climate-related disasters as a TCE in relation to sleep and sleep disparities

In this narrative summary of prior literature related to traumatic childhood experiences, we provide examples from the strongest evidence versus an exhaustive review. Three prior reviews have reported on associations between childhood trauma and sleep over the life course (12, 13, 89). In a systematic review of 20 original research studies that were focused on women, most of the studies reported a significant association between traditionally studied ACEs and sleep disorders during adulthood including sleep apnoea, narcolepsy, nightmare distress, sleep paralysis and psychiatric sleep disorders (13). The authors also suggested that the strengths of associations increased with the frequency and severity of ACEs. More recently, a systematic review of literature published until February 2021 reported that ACEs were associated with poor sleep characteristics, including short sleep duration and long-term sleep problems during adulthood among both men and women (89). Even more recently, literature published up to July 2021 supported linkages between childhood maltreatment and, most commonly, sleep problems during childhood as well as poor sleep quality during adulthood (12). However, climate-related trauma was understudied. To our knowledge, empirical studies among a population of US women are the few that have investigated natural disasters – which are related to climate change – as TCEs in relation to sleep health and sleep health disparities during adulthood (59, 90). These studies reported experience of natural disaster during childhood was associated with multiple sleep dimensions, including short sleep duration and poorer sleep quality (59, 90). Of note, NH-Black and Latina women had two times the prevalence reporting natural disasters during childhood than NH-White women (59). Extrapolating these results with a recent review, impacts of climate change on sleep are currently largely overlooked and understudied but should be considered as a TCE (15). Further, socioeconomic, racial/ethnic and gender disparities in the relationships between climate-related childhood trauma and sleep are plausible, and additional investigation is warranted.

Climate-related disasters as a TCE in relation to depression and anxiety

Several reviews of the literature have elucidated linkages between traumatic childhood experiences and depression during adolescence and adulthood (91–93); however, these studies did not generally include climate change-related trauma during childhood as an adverse childhood experience. Nonetheless, several review studies have demonstrated the detrimental impacts of exposure to climate change events (namely disasters) in relation to mental health with depression and anxiety as two of the most common mental health impacts of climate-related events (16, 65, 67–71, 74–76). We recommend readers refer to the numerous reviews that further describe climate change in relation to mental health, including a review focusing particularly on the impact on children since there is a small yet growing body of literature focusing specifically on the mental health impacts of climate change among children (16). In this article, Burke et al. describe how aforementioned direct, indirect, and vicarious climate change factors impact children's mental health (16). Further, groups experiencing marginalisation are at the highest risk for the mental impacts of climate change, including children, the elderly, women, people (and countries) with low socioeconomic status, outdoor labourers, the racially/ethnically minoritised, immigrants, and people with pre-existing health conditions as well as intersecting marginalised social identities (65, 67, 70, 71, 75). Therefore, inequities across a wide range of populations are poised to increase (15). In fact, Pumariega et al. highlighted the potential widening of racial/ethnic disparities in mental health in the US due to climate change, particularly among children. Recent studies demonstrated that socioeconomically and racially/ethnically minorised children in the US (including Puerto Rico) were disproportionately affected by natural disasters as well as disaster-related stress and mental health outcomes (e.g., depression) (60). Studies also showed long-term consequences of persistent psychological distress and posttraumatic stress after disaster longitudinally among youth and that these mental health adversities persisted longer among minoritised racial/ethnic groups (60). Disparities are likely given that both women and children were identified as some of the most susceptible populations to the mental health impacts of climate change and given the higher burden of both depression and anxiety symptoms among women compared to men (49, 71). Additional studies are needed.

Sleep as a mediator between climate-related disasters as a TCE in relation to depression and anxiety

We identified one study that investigated sleep duration as a mediator between traditionally studied ACEs and depression among young Hispanic and non-Hispanic adults (94). These ACEs did not include natural disaster but results support that sleep is a potential mediating pathway through which TCEs may impact depressive symptoms in young adulthood (94). Regarding climate change as a TCE, data are also sparse. Literature reviews of climate change events rarely mentioned sleep disturbances as consequences beyond as symptoms of adverse mental health outcomes (68, 69, 76). In a review aimed to elucidate a causal pathways framework between climate change and mental health, sleep was not included as a potential mechanism (65) but warrants consideration and investigation due to the shared physical and social environmental determinants of each and because climate change events are related to mental health outcomes for which symptoms include sleep disturbances (69). Further, another review suggested that climate change, in addition to causing physiological stress, may cause reductions in health behaviours like sleep, and additional studies are necessary to identify casual pathways (74) as well as to elucidate socioeconomic, racial/ethnic and gender disparities as potential modifiers of the potentially mediating relationship.

Limitations, future directions and public health implications

Although strengths of the literature include many reviews and a growing body of evidence on the impacts of climate change and sleep on health, there are notable limitations. Importantly, measurement limitations related to subjectively measured sleep and mental health across diverse populations must be addressed concurrently with more frequent investigation/measurement of climate-related trauma as a TCE. No studies, to our knowledge, investigated sleep as a mediator between climate change-related trauma and depression and anxiety. Therefore, robust, prospective data collection across diverse, international populations is necessary to determine causal relationships between climate change-related trauma, sleep and mental health. This approach can further elucidate both biological and social pathways as well as guide intervention efforts.

Despite the limitations, there are important public health implications. While the goal is to eliminate climate change, climate resilience and disaster preparedness efforts are warranted in the meantime (72). These efforts must be equitably distributed by simultaneously addressing 'upstream' or fundamental environmental determinants. For instance, resiliency strategies (e.g., infrastructure development, access to healthcare/psychological services, training of mental health professionals to address climate stressors) to mitigate as well as proactively adapt to climate change is of utmost importance given the potential long-term effects of climate-related TCEs (17), especially in lower-resourced, vulnerable communities (17, 72, 95). Furthermore, greater emphasis should be placed on building coping strategies, self-efficacy, social support and resilience in children to prevent potential long-term impacts of climate-related disaster on sleep and mental health (72, 95–98). To build resiliency in children, education systems should prioritise disaster preparedness programs as well as social and emotional learning and development (72, 98). Further, children should be encouraged to participate in climate change advocacy and mitigation programs, which will foster autonomy in addressing climate change, decrease feelings of helplessness, and contribute to reducing individual-, family and community vulnerability to climate change (98). Long-term participation may also consequentially lead to fewer disasters and thus fewer health consequences related to climate change (98). Additionally, efforts to strengthen collaborations across community and social networks can assist communities in preparing to work together when faced with disaster (95). Lastly and most importantly, it is necessary to address 'upstream' environmental determinants including climate change, structural racism, structural sexism and poverty, which requires multilevel interventions and intentional policies aimed at promoting health equity.

Conclusion

Expected to increase and disproportionately impact socially vulnerable groups, direct and vicarious experiences with climate change events (e.g., natural disasters) are risk factors for poor mental health outcomes such as depression and anxiety over the life course. Sleep is a modifiable contributor to mental health and is inequitably distributed by socioeconomic status, race/ethnicity, and gender groups beginning in early life and persisting over the life course. Sleep likely mediates the climate-related disaster–mental health relationship. However, data are sparse. Climate-related disasters should be included as a TCE in future studies, and both observational and intervention studies that use a health disparities lens are necessary to further illuminate the role of sleep and to address the likely imminent exacerbation of sleep and mental health disparities.

References

1 Chattu VK, Manzar MD, Kumary S, Burman D, Spence DW, Pandi-Perumal SR. The global problem of insufficient sleep and its serious public health implications. Healthcare (Basel). 2018;7(1):1.
2 Institute of Medicine Committee on Sleep M, Research. The national academies collection: reports funded by national institutes of health. In: Colten HR, Altevogt BM, editors. Sleep Disorders and Sleep Deprivation: An Unmet Public Health Problem. Washington, DC: National Academies Press (US) Copyright © 2006, National Academy of Sciences; 2006.
3 Wheaton AG, Jones SE, Cooper AC, Croft JB. Short sleep duration among middle school and high school students – United States, 2015. MMWR Morb Mortal Wkly Rep. 2018;67(3):85–90.
4 Consensus Conference P, Watson NF, Badr MS, Belenky G, Bliwise DL, Buxton OM, et al. Joint consensus statement of the American academy of sleep medicine and sleep research society on the recommended amount of sleep for a healthy adult: methodology and discussion. Sleep. 2015;38(8):1161–83.
5 Johnson DA, Jackson CL, Williams NJ, Alcántara C. Are sleep patterns influenced by race/ethnicity – a marker of relative advantage or disadvantage? Evidence to date. Nat Sci Sleep. 2019;11:79–95.
6 Grandner MA. Sleep, health, and society. Sleep Med Clin. 2022;17(2):117–39.
7 Simonelli G, Petit D, Delage JP, Michaud X, Lavoie MD, Morin CM, et al. Sleep in times of crises: a scoping review in the early days of the COVID-19 crisis. Sleep Med Rev. 2021;60:101545.
8 Leading causes of mortality and health loss at regional, subregional, and country levels in the Region of the Americas, 2000–2019: Pan American Health Organization; 2021. Available from: www.paho.org/en/enlace/leading-causes-death-and-disability#:~:text=Leading%20causes%20of%20disability&text=Back%20and%20neck%20pain%2C%20diabetes,top%205%20causes%20of%20YLDs.
9 Nochaiwong S, Ruengorn C, Thavorn K, Hutton B, Awiphan R, Phosuya C, et al. Global prevalence of mental health issues among the general population during the coronavirus disease-2019 pandemic: a systematic review and meta-analysis. Sci Rep. 2021;11(1):10173.
10 Jahrami HA, Alhaj OA, Humood AM, Alenezi AF, Fekih-Romdhane F, AlRasheed MM, et al. Sleep disturbances during the COVID-19 pandemic: a systematic review, meta-analysis, and meta-regression. Sleep Med Rev. 2022;62:101591.
11 Kessler RC, Chiu WT, Hwang IH, Puac-Polanco V, Sampson NA, Ziobrowski HN, et al. Changes in prevalence of mental illness among US adults during compared with before the COVID-19 pandemic. Psychiatr Clin North Am. 2022;45(1):1–28.
12 Brown SM, Rodriguez KE, Smith AD, Ricker A, Williamson AA. Associations between childhood maltreatment and behavioral sleep disturbances across the lifespan: a systematic review. Sleep Med Rev. 2022;64:101621.
13 Kajeepeta S, Gelaye B, Jackson CL, Williams MA. Adverse childhood experiences are associated with adult sleep disorders: a systematic review. Sleep Med. 2015;16(3):320–30.
14 Baglioni C, Nanovska S, Regen W, Spiegelhalder K, Feige B, Nissen C, et al. Sleep and mental disorders: a meta-analysis of polysomnographic research. Psychol Bull. 2016;142(9):969–90.
15 Bragazzi NL, Garbarino S, Puce L, Trompetto C, Marinelli L, Currà A, et al. Planetary sleep medicine: studying sleep at the individual, population, and planetary level. Front Public Health. 2022;10.
16 Burke SEL, Sanson AV, Van Hoorn J. The psychological effects of climate change on children. Curr Psychiatry Rep. 2018;20(5):35.

17 Watts N, Amann M, Ayeb-Karlsson S, Belesova K, Bouley T, Boykoff M, et al. The Lancet countdown on health and climate change: from 25 years of inaction to a global transformation for public health. Lancet. 2018;391(10120):581–630.

18 Duran DG, Pérez-Stable EJ. Novel approaches to advance minority health and health disparities research. Am J Public Health. 2019;109(S1):S8–S10.

19 Centers for Disease Control and Prevention. Advancing Health Equity in Chronic Disease Prevention and Management: National Center for Chronic Disease Prevention and Health Promotion; 2022. Available from: www.cdc.gov/chronicdisease/healthequity/index.htm.

20 Billings ME, Cohen RT, Baldwin CM, Johnson DA, Palen BN, Parthasarathy S, et al. Disparities in sleep health and potential intervention models: a focused review. Chest. 2021;159(3):1232–40.

21 Jackson CL, Powell-Wiley TM, Gaston SA, Andrews MR, Tamura K, Ramos A. Racial/ethnic disparities in sleep health and potential interventions among women in the United States. J Womens Health (Larchmt). 2020;29(3):435–42.

22 Jackson CL, Walker JR, Brown MK, Das R, Jones NL. A workshop report on the causes and consequences of sleep health disparities. Sleep. 2020;43(8).

23 Ahn S, Lobo JM, Logan JG, Kang H, Kwon Y, Sohn MW. A scoping review of racial/ethnic disparities in sleep. Sleep Med. 2021;81:169–79.

24 Paul KN, Turek FW, Kryger MH. Influence of sex on sleep regulatory mechanisms. J Womens Health (Larchmt). 2008;17(7):1201–8.

25 Butler ES, McGlinchey E, Juster RP. Sexual and gender minority sleep: a narrative review and suggestions for future research. J Sleep Res. 2020;29(1):e12928.

26 Etindele Sosso FA, Holmes SD, Weinstein AA. Influence of socioeconomic status on objective sleep measurement: a systematic review and meta-analysis of actigraphy studies. Sleep Health. 2021;7(4):417–28.

27 Etindele Sosso FA. Measuring sleep health disparities with polysomnography: a systematic review of preliminary findings. Clocks Sleep. 2022;4(1):80–7.

28 Etindele Sosso FA, Kreidlmayer M, Pearson D, Bendaoud I. Towards a socioeconomic model of sleep health among the Canadian population: a systematic review of the relationship between age, income, employment, education, social class, socioeconomic status and sleep disparities. Eur J Investig Health Psychol Educ. 2022;12(8):1143–67.

29 Smith JP, Hardy ST, Hale LE, Gazmararian JA. Racial disparities and sleep among preschool aged children: a systematic review. Sleep Health. 2019;5(1):49–57.

30 Chattu VK, Chattu SK, Spence DW, Manzar MD, Burman D, Pandi-Perumal SR. Do disparities in sleep duration among racial and ethnic minorities contribute to differences in disease prevalence? J Racial Ethn Health Disparities. 2019;6(6):1053–61.

31 Guglielmo D, Gazmararian JA, Chung J, Rogers AE, Hale L. Racial/ethnic sleep disparities in US school-aged children and adolescents: a review of the literature. Sleep Health. 2018;4(1):68–80.

32 Liu J, Ji X, Pitt S, Wang G, Rovit E, Lipman T, et al. Childhood sleep: physical, cognitive, and behavioral consequences and implications. World J Pediatr. 2022:1–11.

33. Dejenie TA, G/Medhin MT, Admasu FT, Adella GA, Enyew EF, Kifle ZD, et al. Impact of objectively-measured sleep duration on cardiometabolic health: a systematic review of recent evidence. Front Endocrinol (Lausanne). 2022;13:1064969.

34 Scott AJ, Webb TL, Martyn-St James M, Rowse G, Weich S. Improving sleep quality leads to better mental health: a meta-analysis of randomised controlled trials. Sleep Med Rev. 2021;60:101556.

35 Lam LT, Lam MK. Sleep disorders in early childhood and the development of mental health problems in adolescents: a systematic review of longitudinal and prospective studies. Int J Environ Res Public Health. 2021;18(22).

36 Roenneberg T, Foster RG, Klerman EB. The circadian system, sleep, and the health/disease balance: a conceptual review. J Sleep Res. 2022;31(4):e13621.

37 Laposky AD, Van Cauter E, Diez-Roux AV. Reducing health disparities: the role of sleep deficiency and sleep disorders. Sleep Med. 2016;18:3–6.

38 Jackson CL, Redline S, Emmons KM. Sleep as a potential fundamental contributor to disparities in cardiovascular health. Annu Rev Public Health. 2015;36:417–40.

39 Allen J, Balfour R, Bell R, Marmot M. Social determinants of mental health. Int Rev Psychiatry. 2014;26(4):392–407.

40 Miconi D, Li ZY, Frounfelker RL, Santavicca T, Cénat JM, Venkatesh V, et al. Ethno-cultural disparities in mental health during the COVID-19 pandemic: a cross-sectional study on the impact of exposure to the virus and COVID-19-related discrimination and stigma on mental health across ethno-cultural groups in Quebec (Canada). BJPsych Open. 2021;7(1):e14.

41 Tiwari SK, Wang J. Ethnic differences in mental health service use among White, Chinese, South Asian and South East Asian populations living in Canada. Soc Psychiatry Psychiatr Epidemiol. 2008;43(11):866–71.

42 Alegria M, Perez DJ, Williams S. The role of public policies in reducing mental health status disparities for people of color. Health Aff (Millwood). 2003;22(5):51–64.

43 Phimphasone-Brady P, Page CE, Ali DA, Haller HC, Duffy KA. Racial and ethnic disparities in women's mental health: a narrative synthesis of systematic reviews and meta-analyses of U.S. samples. Fertil Steril. 2023;199(3):364–74.

44 Himle JA, Baser RE, Taylor RJ, Campbell RD, Jackson JS. Anxiety disorders among African Americans, blacks of Caribbean descent, and non-Hispanic whites in the United States. J Anxiety Disord. 2009;23(5):578–90.

45 Lemonious T, Codner M, Pluhar E. A review on the disparities in the identification and assessment of depression in Black adolescents and young adults. How can clinicians help to close the gap? Curr Opin Pediatr. 2022;34(4):313–19.

46 Clark TT, Salas-Wright CP, Vaughn MG, Whitfield KE. Everyday discrimination and mood and substance use disorders: a latent profile analysis with African Americans and Caribbean Blacks. Addict Behav. 2015;40:119–25.

47 Stone D, Mack K, Qualters J. Notes from the field: recent changes in suicide rates, by race and ethnicity and age group – United States, 2021. MMWR Morb Mortal Wkly Rep. 2023;72(6):160–2.

48 Perez NB, Lanier Y, Squires A. Inequities along the depression care cascade in African American women: an integrative review. Issues Ment Health Nurs. 2021;42(8):720–9.

49 Kundakovic M, Rocks D. Sex hormone fluctuation and increased female risk for depression and anxiety disorders: from clinical evidence to molecular mechanisms. Front Neuroendocrinol. 2022;66:101010.

50 Li SH, Graham BM. Why are women so vulnerable to anxiety, trauma-related and stress-related disorders? The potential role of sex hormones. Lancet Psychiatry. 2017;4(1):73–82.

51 Lasiuk GC, Hegadoren KM. The effects of estradiol on central serotonergic systems and its relationship to mood in women. Biol Res Nurs. 2007;9(2):147–60.

52 Altemus M, Sarvaiya N, Neill Epperson C. Sex differences in anxiety and depression clinical perspectives. Front Neuroendocrinol. 2014;35(3):320–30.

53 Lin J, Zou L, Lin W, Becker B, Yeung A, Cuijpers P, et al. Does gender role explain a high risk of depression? A meta-analytic review of 40 years of evidence. J Affect Disord. 2021;294:261–78.

54 World Health Organization. Social Determinants of Health 2023. Available from: www.who.int/health-topics/social-determinants-of-health#tab=tab_1.

55 Jackson CL, Gaston SA. The impact of environmental exposures on sleep. In: Grandner MA, editor. Sleep and Health. Netherlands: Elsevier, Academic Press; 2019. p. 85–103.

56 Billings ME, Hale L, Johnson DA. Physical and social environment relationship with sleep health and disorders. Chest. 2020;157(5):1304–12.

57 Bailey ZD, Krieger N, Agenor M, Graves J, Linos N, Bassett MT. Structural racism and health inequities in the USA: evidence and interventions. Lancet. 2017;389(10077):1453–63.

58 Homan P, Brown TH, King B. Structural intersectionality as a new direction for health disparities research. J Health Soc Behav. 2021;62(3):350–70.

59 Gaston SA, McWhorter KL, Parks CG, D'Aloisio AA, Rojo-Wissar DM, Sandler DP, et al. Racial/ethnic disparities in the relationship between traumatic childhood experiences and suboptimal sleep dimensions among adult women: findings from the sister study. Int J Behav Med. 2021;28(1):116–29.

60 Pumariega AJ, Jo Y, Beck B, Rahmani M. Trauma and US minority children and youth. Curr Psychiatry Rep. 2022;24(4):285–95.

61 Kendler KS, Kessler RC, Neale MC, Heath AC, Eaves LJ. The prediction of major depression in women: toward an integrated etiologic model. Am J Psychiatry. 1993;150(8):1139–48.

62 Cutler SE, Nolen-Hoeksema S. Accounting for sex differences in depression through female victimization: childhood sexual abuse. Sex Roles. 1991;24(7):425–38.

63 Felitti VJ, Anda RF, Nordenberg D, Williamson DF, Spitz AM, Edwards V, et al. Relationship of childhood abuse and household dysfunction to many of the leading causes of death in adults: the adverse childhood experiences (ACE) study. Am J Prev Med. 1998;14(4):245–58.

64 Bellis MA, Hughes K, Ford K, Ramos Rodriguez G, Sethi D, Passmore J. Life course health consequences and associated annual costs of adverse childhood experiences across Europe and North America: a systematic review and meta-analysis. Lancet Public Health. 2019;4(10):e517–28.

65 Berry HL, Bowen K, Kjellstrom T. Climate change and mental health: a causal pathways framework. Int J Public Health. 2010;55(2):123–32.

66 Hidalgo J, Baez AA. Natural disasters. Crit Care Clin. 2019;35(4):591–607.

67 Palinkas LA, Wong M. Global climate change and mental health. Curr Opin Psychol. 2020;32:12–16.

68 Doherty TJ, Clayton S. The psychological impacts of global climate change. Am Psychol. 2011;66:265–76.

69 Trombley J, Chalupka S, Anderko L. Climate change and mental health. AJN Am J Nurs. 2017;117(4):44–52.

70 Fritze JG, Blashki GA, Burke S, Wiseman J. Hope, despair and transformation: climate change and the promotion of mental health and wellbeing. Int J Mental Health Syst. 2008;2(1):13.

71 Dodgen D, Donato D, Kelly N, La Greca A, Morganstein J, Reser J, Ruzek J, Schweitzer S, Shimamoto MM, Thigpen Tart K, Ursano R. Ch. 8: Mental Health and Well-Being. The Impacts of Climate Change on Human Health in the United States: A Scientific Assessment. Washington, DC: U.S. Global Change Research Program; 2016, p. 217–46. http://dx.doi. org/10.7930/J0TX3C9H

72 Sheffield PE, Landrigan PJ. Global climate change and children's health: threats and strategies for prevention. Environ Health Perspect. 2011;119(3):291–8.

73 Environmental Protection Agency. Climate Change and Social Vulnerability in the United States: A Focus on Six Impacts. U.S. Environmental Protection Agency; 2021. Contract No.: EPA 430-R-21-003. www.epa.gov/cira/social-vulnerability-report

74 Obradovich N, Migliorini R, Paulus MP, Rahwan I. Empirical evidence of mental health risks posed by climate change. Proc Natl Acad Sci. 2018;115(43):10953–8.

75 Hayes K, Blashki G, Wiseman J, Burke S, Reifels L. Climate change and mental health: risks, impacts and priority actions. Int J Ment Health Syst. 2018;12(1):28.

76 Clayton S, Manning CM, Hodge C. Beyond Storms & Droughts: The Psychological Impacts of Climate Change. Washington, DC: American Psychological Association and ecoAmerica; 2014.

77 Rifkin DI, Long MW, Perry MJ. Climate change and sleep: a systematic review of the literature and conceptual framework. Sleep Med Rev. 2018;42:3–9.

78 Berens AE, Jensen SKG, Nelson CA. Biological embedding of childhood adversity: from physiological mechanisms to clinical implications. BMC Med. 2017;15(1):135.

79 Heim C, Nemeroff CB. The role of childhood trauma in the neurobiology of mood and anxiety disorders: preclinical and clinical studies. Biol Psychiatry. 2001;49(12):1023–39.

80 Ridout KK, Khan M, Ridout SJ. Adverse childhood experiences run deep: toxic early life stress, telomeres, and mitochondrial DNA copy number, the biological markers of cumulative stress. Bioessays. 2018;40(9):e1800077.

81 McEwen BS, Karatsoreos IN. Sleep deprivation and circadian disruption stress, allostasis, and allostatic load. Sleep Med Clin. 2022;17(2):253–62.

82 Herzog JI, Schmahl C. Adverse childhood experiences and the consequences on neurobiological, psychosocial, and somatic conditions across the lifespan. Front Psychiatry. 2018;9:420.

83 Jaworska-Andryszewska P, Rybakowski JK. Childhood trauma in mood disorders: neurobiological mechanisms and implications for treatment. Pharmacol Rep. 2019;71(1):112–20.

84 Jiang S, Postovit L, Cattaneo A, Binder EB, Aitchison KJ. Epigenetic modifications in stress response genes associated with childhood trauma. Front Psychiatry. 2019;10:808.

85 Sammallahti S, Koopman-Verhoeff ME, Binter A-C, Mulder RH, Cabré-Riera A, Kvist T, et al. Longitudinal associations of DNA methylation and sleep in children: a meta-analysis. Clin Epigenetics. 2022;14(1):83.

86 Dumornay NM, Lebois LA, Ressler KJ, Harnett NG. Racial disparities in adversity during childhood and the false appearance of race-related differences in brain structure. Am J Psychiat. 2023;180(2):127–38.

87 Barandouzi ZA, Starkweather AR, Henderson WA, Gyamfi A, Cong XS. Altered composition of gut microbiota in depression: a systematic review. Front Psychiatry. 2020;11:541.

88 Wang Z, Wang Z, Lu T, Chen W, Yan W, Yuan K, et al. The microbiota-gut-brain axis in sleep disorders. Sleep Med Rev. 2022;65:101691.

89 Vadukapuram R, Shah K, Ashraf S, Srinivas S, Elshokiry AB, Trivedi C, et al. Adverse childhood experiences and their impact on sleep in adults: a systematic review. J Nerv Ment Dis. 2022;210(6):397–410.

90 McWhorter KL, Parks CG, D'Aloisio AA, Rojo-Wissar DM, Sandler DP, Jackson CL. Traumatic childhood experiences and multiple dimensions of poor sleep among adult women. Sleep. 2019;42(8).

91 Assini-Meytin LC, Fix RL, Green KM, Nair R, Letourneau EJ. Adverse childhood experiences, mental health, and risk behaviors in adulthood: exploring sex, racial, and ethnic group differences in a nationally representative sample. J Child Adolesc Trauma. 2022;15(3):833–45.

92 Sahle BW, Reavley NJ, Li W, Morgan AJ, Yap MBH, Reupert A, et al. The association between adverse childhood experiences and common mental disorders and suicidality: an umbrella review of systematic reviews and meta-analyses. Eur Child Adolesc Psychiatry. 2022;31(10):1489–99.

93 Elmore AL, Crouch E. Anxiety, depression, and adverse childhood experiences: an update on risks and protective factors among children and youth. Acad Pediatr. 2023;23(4):720–1.

94 O'Neill RM, Cundiff JM, Wendel CJ, Schmidt AT, Cribbet MR. An examination of sleep as a mediator of the relationship between childhood adversity and depression in hispanic and non-hispanic young adults. Behav Med. 2022:1–12.

95 Trombley J, Chalupka S, Anderko L. Climate change and mental health. Am J Nurs. 2017;117(4).

96 Grolnick WS, Schonfeld DJ, Schreiber M, Cohen J, Cole V, Jaycox L, et al. Improving adjustment and resilience in children following a disaster: addressing research challenges. Am Psychol. 2018;73(3):215–29.

97 Hayes K, Poland B. Addressing mental health in a changing climate: incorporating mental health indicators into climate change and health vulnerability and adaptation assessments. Int J Environ Res Public Health. 2018;15(9).

98 Seddighi H, Yousefzadeh S, López López M, Sajjadi H. Preparing children for climate-related disasters. BMJ Paediatrics Open. 2020;4(1):e000833.

12

THE PAST, PRESENT AND FUTURE OF SLEEP MONITORING TECHNOLOGIES

Hannah Scott and Bastien Lechat

The objective measurement of sleep has a long history. Since the first electroencephalography (EEG) recordings of human sleep in 1929 (1), the past century of sleep research has involved developing and applying objective techniques to understand the normal and abnormal physiological changes that occur with sleep. Today, sleep monitoring technologies are an integral part of sleep research and clinical practice. In fact, the development of actigraphy-based devices has led to the discovery of most of what is known about objectively assessed sleep in the home environment. Sleep monitoring technologies are also used in routine clinical care to assist in the diagnosis of some sleep disorders. Rapid advances in the capabilities of these technologies underpin major new research insights into sleep and health, present new opportunities and challenges for clinical practice and have prompted a wave of consumers using these devices to better understand their sleep.

This chapter describes the most common and emerging techniques used in wearable and non-wearable trackers for estimating sleep, including their major strengths and limitations for multi-night sleep monitoring. Next, the utility of sleep monitoring technologies is described for researchers, clinicians and the general population to better monitor and manage sleep to support overall health. Lastly, predictions are made about the future of sleep monitoring technologies and how they may be applied to actively address and personalise the management of sleep and its disorders.

Sleep assessment techniques

The last century of sleep research has seen the development of new objective sleep monitoring techniques and the simplification and miniaturisation

DOI: 10.4324/9781003296966-12

of traditional techniques for general use outside of the sleep laboratory. Figure 12.1 outlines a brief chronological history of the major techniques and their application in objective sleep monitoring devices. These techniques are described below, to serve as a guide for understanding how sleep monitoring technologies acquire sleep-related data. The algorithms that are applied to the data to score sleep are also worthy of consideration and are discussed elsewhere (2).

Techniques implemented in wearable devices

Actigraphy

Actigraphy measures motor activity via an accelerometer device to measure acceleration (speed of movement) along three axes (x, y, z) many times per second, with these values typically averaged over a 30-second epoch (3, 4). When worn continuously (often on the wrist), sleep can be estimated as the absence of movement via numerous sleep-wake scoring algorithms. Actigraphy was first used to estimate sleep in 1982 (5) and became a hugely popular technique due to its non-invasive nature and ability to provide multi-night monitoring of sleep patterns. Even in modern sleep monitoring devices, actigraphy is often used as the foundational technique for sleep-wake assessment.

The popularity of actigraphy-based devices for estimating sleep is largely due to its practicality for long-term sleep assessments in the home environment (6). Actigraphy-based devices are non-invasive, comfortable to wear and simple for individuals to operate (therefore, often have high compliance), low

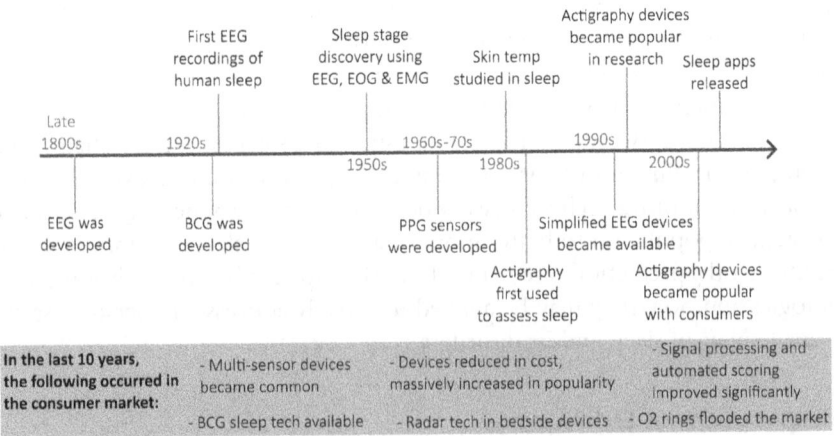

FIGURE 12.1 A brief chronological history of the major techniques applied in objective sleep monitoring devices.

cost compared to traditional sleep assessments, providing continuous monitoring over a long period. These features enable the assessment of long-term trends in sleep and activity patterns in the naturalistic environment where such assessment would otherwise be infeasible.

While actigraphy devices have many advantages, there are some inherent limitations to consider. Given that actigraphy-based devices rely upon accelerometers to measure movement which can be affected by a variety of factors, their accuracy is limited (7, 8). Individuals often lie still in bed while awake: a behaviour that is often scored as sleep by these devices. Additionally, while many actigraphy-based devices incorporate a button for individuals to signal in/out of bed, which is used to varying degrees of compliance, the in-bed period is difficult to discern from accelerometer data, which further impairs the accuracy of sleep estimates. Scoring algorithms relying solely upon accelerometry signals are also unable to differentiate between sleep stages as the motor activity levels detectable via these devices are largely comparable across stages (9). With actigraphy devices that require being removed/reapplied (e.g., due to battery charging or avoiding contact with water), data loss can be a significant limitation due to individuals not wearing the device.

Photoplethysmography (PPG)

PPG is a technique used to measure blood volume changes in the microvascular bed of skin tissue (10, 11). This technique involves shining a light onto the skin and measuring the amount of light that is absorbed and reflected. When blood flows through the vessels in the skin, it absorbs more light, leading to a decrease in the amount of light reflected back to the sensor. This change in the amount of light reflected is used to calculate the blood volume in the tissue, which in turn can be used to determine heart rate, oxygen saturation, and other physiological parameters. Given that heart rate variability changes with sleep (higher during sleep than wake, higher during NREM sleep than REM sleep) (12, 13), PPG is useful for estimating transitions between sleep and wake and for distinguishing sleep stages. Oxygen saturation levels are also useful for estimating sleep stages (14) and sleep-disordered breathing (15, 16). Accordingly, many wearable sleep monitors incorporate PPG signals.

PPG requires a light source and a light detector. These sensors are small and inexpensive, meaning that PPG is a cost-effective technique that can be incorporated into wearable monitors relatively easily. Like actigraphy, PPG is also a non-invasive method that can provide continuous monitoring in comfortable and simple-to-operate devices. However, the quality of PPG signals is affected by multiple factors, including ambient light, contact with skin and movement, which can lead to measurement error and inaccurate results (17, 18). Skin pigmentation also affects PPG signals because different skin pigmentations reflect light differently, influencing the derivation of heart rate and other parameters.

Skin temperature sensor

While temperature is not a signal used in traditional sleep assessments, their inclusion in sleep monitoring technologies serves a useful purpose. Acute changes occur in distal temperature around the onset/offset of sleep (19), and as such, skin temperature signals can be used to support estimations of sleep-wake transitions (20–22). In addition, core body temperature shows a circadian rhythm, which can be estimated from skin temperature data given sufficient signal quality (23, 24). This technique is therefore promising for potentially improving sleep assessment accuracy and for providing supplementary information about circadian rhythms.

Small, inexpensive temperature sensors can be used to continuously derive skin temperature, meaning that this technique is practical for incorporation into many wearable monitors. The potential for accurate estimation of circadian rhythm parameters via skin temperature is also a major advantage, given that other physiological techniques are largely incapable of providing accurate estimates of this major biological determinant of sleep. Importantly, skin temperature is influenced by multiple factors other than circadian-related factors, including the blood flow in the underlying tissue. This means that skin temperature may not provide accurate estimations of body temperature rhythms, especially during physical activity or when the body is experiencing changes in blood flow due to illness or times of physiological stress (25, 26). Similarly, signal quality is affected by ambient temperature, humidity, airflow and sweat. Therefore, a single skin temperature sensor without an associated ambient environmental sensor and/or multiple other temperature sensors on the body to compare currently has limited accuracy in uncontrolled environments. Nonetheless, as advancements are made in signal processing with the aid of techniques like machine learning, these limitations may be overcome in the future.

Electroencephalography (EEG)

A signal from an EEG electrode represents the postsynaptic potential from millions of neurons located in close proximity to the electrode, largely from cortical neurons near the scalp. Since the early recordings of human brain activity and the discovery of rapid eye movement (REM) sleep and non-rapid eye movement (NREM) sleep stages in 1957 (27), EEG remains a fundamental component to the gold-standard method of assessing sleep, polysomnography (28). Today, simplified EEG is incorporated into multiple sleep monitors to provide more in-depth assessments of sleep stages and sleep quality.

Given that the gold-standard sleep assessment relies considerably upon EEG, devices that include EEG are inherently 'accurate' for estimating polysomnographically derived sleep (29–31). With high sampling rates, EEG-based devices can be used to assess sleep stages, detect many sleep-specific EEG features (e.g., k-complexes, sleep spindles, EEG slow waves) and support the assessment of many sleep disorders. The development of more practical electrodes (e.g., dry electrodes, printable sensors), simpler EEG acquisition systems and automated sleep scoring algorithms (32, 33) also mean that many of the traditional limitations of EEG have been largely overcome. However, considerable limitations remain, namely that EEG-based devices must be set up every night and are typically less comfortable and more expensive than other sleep monitoring technologies. Collectively, these limitations affect the feasibility of, and individual's willingness to use, EEG-based devices over multiple nights.

Given that EEG-based devices score the data into traditional sleep stages, the limitations of traditional sleep scoring are also relevant here. Since the brain activity is scored into 30-second epochs of discrete sleep stages, only gross EEG features are captured with much poorer time and frequency resolution than is available within the data and largely ignoring that wake and sleep are continuous and dynamic states (34, 35). Currently implemented methods and technologies largely do not address this problem. Top-down sleep scoring approaches of important sleep signal features are needed to overcome this limitation and to better define sleep disruption (36).

Other techniques incorporated into wearable sleep monitoring devices

- *Electrodermal activity (EDA)* can be assessed via wearable sensors in some sleep monitors (37–40). Aside from EDA parameters being useful markers of physiological activation (since skin conductance is influenced by sweat which is controlled via the sympathetic nervous system), there are also sleep stage-dependent changes in skin conductivity. In particular, this technique can be used to differentiate N2 and N3 stages from other sleep stages (39).
- *Electrooculography (EOG)* is a technique used to measure the electrical potential difference between the front and the back of the eye (28). Traditionally implemented during polysomnographic sleep studies via an electrode placed near each eye, the development of dry electrodes and other technologies has enabled the use of EOG in head-worn devices and eye masks particularly to assist in the detection of REM sleep (41, 42).
- *Electromyography (EMG)* is a technique used to measure and record the electrical activity produced by skeletal muscles (28). This technique is implemented as part of polysomnography but is also being incorporated into devices worn on the chin to assist in the assessment of breathing during sleep and sleep bruxism (43–46).
- *Electrocardiography (ECG)* is a technique used to measure and record the electrical activity of the heart (28). This technique is being incorporated into multiple wearable devices via printed electrodes placed on the chest and wearable fabrics (i.e., shirts) (47–49).

- *Respiratory inductance plethysmography (RIP)* is a non-invasive technique for monitoring breathing patterns. Typically administered belts placed around the abdomen, the method can measure the movement of the chest and abdominal wall that occur during inspiration and expiration during sleep (50).

Techniques implemented in non-wearable devices

Ballistography

Ballistography is a highly accurate set of techniques for capturing forces generated by the body, such as the body's motion (displacement, velocity or acceleration) from limb movements, respiration or from the cardiac ejection of blood (specifically termed ballistocardiography) (51, 52). When blood moves downward through the descending aorta, the body subtly recoils upward, producing a movement with every heartbeat. Using a pneumatic sensor placed under the mattress, this downward motion, along with other gross body movements and respiratory fluctuations, is detectable as a pressure change with their own characteristic signatures. Some devices use this technique to monitor sleep stages, heart rate, and respiratory fluctuations without any contact with the user (53–55). Ballistography is also highly sensitive for assessing heart rate, sometimes more so than other techniques used in wearable devices.

Radar and sonar technology

Doppler technology is a type of technology that uses the Doppler effect to measure the velocity of objects. The Doppler effect is the change in frequency or wavelength of a wave in relation to an observer who is moving relative to the source of the wave (56). For sleep assessment, this technique can be employed in bedside devices that emit either radio waves (radar) or inaudible sound (sonar) to measure their reflections and derive movement in the environment. These techniques are capable of detecting large body movements and are even sensitive enough to distinguish subtle movements due to respiration and heart rate (57–59). Therefore, these techniques provide contactless methods of estimating sleep stages and breathing during sleep. Additionally, sonar technology is being applied in smartphones (58, 59), meaning that the technique can be administered via equipment that users already own without having to purchase new devices. However, while sonar emits largely inaudible sound, some components can be audible, especially when administered from smartphones with poor audio quality.

Advantages/disadvantages of non-wearables

Major strengths of non-wearable devices are comfort and they are generally more convenient for multi-night recording because they require minimal user effort after initial setup, since they rarely need any adjustment, and are stored in a convenient location. In addition, these techniques can acquire information from the whole body, not just a single site. However, once set up, some devices have limited mobility and are therefore cumbersome for individuals who regularly sleep elsewhere (e.g., while travelling). Additionally, these techniques can be less effective when bedpartners are present as they can interfere with the signal being collected.

Techniques to assess environmental factors

The sleeping environment can significantly impact sleep. Sleep disruption can sometimes be attributed to exogenous factors such as noise, light, temperature and other comfort-related factors. While these factors are typically carefully controlled in a laboratory setting to promote sleep, it is crucial to consider and assess their impact in the less controlled home environment to properly characterise the sleep disruption.

Many sleep monitoring technologies are starting to include sensors that assess environmental factors. While these sensors are often incorporated into sleep technology to infer endogenous factors impacting sleep (e.g., circadian timing, breathing disturbances), their use to assess exogenous factors will become increasingly important through the transition to home-based sleep diagnostics. Techniques that are currently incorporated into some sleep monitoring technologies to assess two influential environmental factors are described below.

Ambient light

Some wearable and non-wearable sleep monitors include light sensors to measure ambient light levels in the environment (60). In wrist-worn sleep trackers, photodiode light sensors are typically used, which can measure the intensity of light over a wide range of wavelengths. Smartphones also include photodiode light sensors as a means of automatically adjusting screen brightness to ensure readability under changing ambient light conditions, but this also clearly has applications for the assessment of sleep.

Ambient light data can serve multiple purposes. One purpose is to assist in defining the sleep opportunity by assessing lights out/light on times and any out-of-bed times during the night (61). This is particularly useful in actigraphy-based devices, which are inadequate for determining when the individual is in bed and attempting sleep. Another purpose to infer circadian timing from light intensity data (62), since light is the strongest exogenous influencer of circadian rhythms. However, the most common ambient light sensors typically detect photopic illuminance rather than melanopic illuminance (the light wavelengths that the circadian system is most responsive to) (63). Additionally, for individuals who are not entrained to (in alignment with) their light-dark schedule (i.e., shift workers), the estimation of circadian timing solely from ambient light data will have limited accuracy.

Sound

Some sleep monitors also include microphones to assess the ambient noise levels in the environment, as well as sounds made by the user during sleep. Most commonly, microphones from smartphones are used for this purpose (64–66), which can assess a wide range of noise characteristics, including the intensity (sound pressure level), frequency and duration of the noise. Several sleep monitoring apps use the smartphone's microphone to detect snoring and apnoeic events, which can be differentiated from other noise types due to their specific acoustic characteristics (67, 68), and infer the presence and severity of sleep-disordered breathing. While uncommon, microphones in sleep monitoring devices can also be used to assess environmental noise during the sleep period. Given that environmental noise can adversely affect sleep via fragmentation, reduced sleep continuity and reduced total sleep time (69–71), its assessment as a potential sleep-disturbing factor is warranted to help determine the cause of sleep disruptions.

Validation of current sleep monitoring technologies

The question arises as to whether current sleep monitoring technology is accurate enough to provide valuable data for its intended uses. Many studies have been conducted to assess the accuracy of these devices in comparison to polysomnography (72–77), which is considered the gold standard for sleep studies. Two main findings have emerged from prior validation studies. First, devices that incorporate multiple sensors tend to assess sleep stages more accurately than single-sensor devices, with the exception of EEG-based devices, which are typically the most accurate devices (9, 31, 78–80). Second, the accuracy of sleep monitoring devices can vary substantially under certain conditions, particularly the algorithms employed and the characteristics of the sample population (such as age and sleep quality) (8, 9, 81, 82). Such findings are critical to consider when selecting sleep monitoring technologies for a given research or clinical purpose.

Notably, prior validation studies have been largely conducted in the controlled sleep laboratory with young, healthy people who sleep well. As such, the accuracy of these technologies in the uncontrolled home environment with people who have poor sleep or sleep disorders, who are often the intended target group, is limited. The rapid speed at which the sleep monitoring technology market is evolving means that thorough evaluation of the accuracy of these devices under their intended conditions is difficult to achieve. However, this is a critical goal to ensure that these technologies are suitable and that the data they collect is reliable (see (6) for a review on these issues).

While sleep monitoring devices are not as accurate as polysomnography, which also has a certain degree of variability in its results (i.e., inter- and intra-scorer variability) (83), it is not yet clear whether they are accurate enough for the proposed uses. Very few studies have established a priori benchmarks to determine the minimum level of accuracy required for a sleep device to be suitable for a specific application (6, 84). Some have suggested that researchers and clinicians should consider the findings of validation studies to determine whether a particular sleep monitoring device is appropriate for their intended use. Others have argued that a more effective approach would be to conduct proof-of-concept studies to assess the feasibility and effectiveness of the technology in enabling the intended uses and providing non-inferior care to patients compared to traditional practices. Ultimately, a combination of proof-of-concept trials and adequately designed and powered randomised-controlled trials are needed to demonstrate the level of patient and clinician acceptance, device accuracy and clinical utility, effectiveness, and cost-effectiveness necessary to support the evidence-based adoption and widespread use of sleep monitoring devices in clinical practice.

Application of sleep monitoring technologies

Sleep monitoring technologies have been used to monitor sleep patterns for a variety of purposes broadly related to characterising multi-night sleep patterns. It is important to note that most sleep monitoring technologies are not medical devices and are therefore unsuitable for medical-related purposes. This said, a growing number of sleep technologies available to consumers are seeking regulatory approval as medical devices. These devices can be highly useful for general purposes as well as the management of sleep disorders in multiple ways. Some major applications of sleep monitoring technologies are discussed below.

Management of sleep disorders

In 2018, a task force commissioned by the American Academy of Sleep Medicine (AASM) provided clinical practice guidelines on the appropriate use of actigraphy-based devices for the management of sleep disorders (85). This task force conducted a systematic review of the available literature to compare data from actigraphy to polysomnography and sleep diaries in people with various sleep disorders (86). Their review found that actigraphy devices provide unique and clinically useful information for adult and paediatric patients with suspected or diagnosed insomnia, circadian rhythm sleep-wake disorders, sleep-disordered breathing, central disorders of hypersomnolence and adults with insufficient sleep syndrome. These devices, however, were found to be unreliable for detecting periodic limb movements and were therefore not recommended for people with periodic limb movement disorder. Current and potential uses of sleep monitoring devices for the management of the two most common sleep disorders are described below.

Insomnia

With respect to insomnia, actigraphy data can be used to aid with making differential diagnoses and for instances where objective estimates of sleep may be important for clinical decision-making (85), for example when people are non-responsive to cognitive behavioural therapy for insomnia, people request an increase in sleep medication dose, and/or the validity of patient reporting is a concern. In practice, some clinicians also use patient data from actigraphy devices as part of the patient's education about 'normal' sleep patterns.

While the diagnosis of insomnia is necessarily reliant upon subjective measures, objective sleep monitoring may be useful for further characterising the nature and severity of sleep disruption. For example, data from sleep monitoring devices could be used to assess for insomnia phenotypes. One phenotype is patients with 'objectively short sleep', defined by single-night polysomnographic sleep studies that indicate ≤6 hours of sleep or sleep efficiency (percent of time spent in bed asleep) ≤85% (87). This phenotype may represent a more biologically severe form of insomnia (88), since there is some evidence (albeit inconsistent) to support that insomnia with objective short sleep is associated with increased risks of hypertension (89), depression (90), and mortality (91), as well as a blunted response to non-pharmacological treatment (87). Another proposed insomnia phenotype that could benefit from objective sleep assessment are patients who experience high levels of sleep-wake state discrepancy (92). Sometimes referred to as sleep-state misperception or paradoxical insomnia, some insomnia patients present with a substantial mismatch between their self-reported sleep duration and the sleep duration determined from concurrent sleep studies (93). While research to demonstrate the clinical utility of assessing and differentially treating these phenotypes is ongoing, the objective assessment of sleep via sleep trackers to detect phenotypes of insomnia may be useful for implementation into clinical practice.

Another potential case for sleep monitoring technologies in the management of insomnia is to facilitate treatment administration. For example, sleep monitoring technologies are being developed to administer Intensive Sleep Retraining outside the laboratory setting (94, 95). Intensive Sleep Retraining is a brief, 24-hour behavioural treatment that pivotal clinical trials found was just as efficacious as four weeks of stimulus control therapy. The biggest barrier to the implementation of this treatment in clinical practice was the reliance on traditional high-cost sleep recordings to inform when to wake patients after they fall asleep during the procedure (96). Sleep monitoring technology is enabling the routine administration of this efficacious yet otherwise impractical technique outside of the laboratory environment (97, 98). Another example is to support the administration of digital treatment programs for insomnia

(99–101). A pilot study recently demonstrated the feasibility of an app that delivered Sleep Restriction Therapy which was supported by sleep monitoring devices (102). These devices were used to assist and adjust the titration of time in bed every three days, with study participants responding favourably to this use of sleep technologies. Although there is potential to cause distress by providing feedback about sleep to patients with insomnia (known as orthosomnia (103, 104)), there is also a clear willingness by some patients to engage with sleep monitoring devices, which may benefit insomnia management if implemented in an appropriate way. While more research is required, these applications demonstrate the potential clinical utility of sleep monitoring technologies for the treatment of insomnia, the success of which may reduce the burden on healthcare systems that are struggling to meet the demand for insomnia treatment services and improve patient care.

Obstructive sleep apnoea

With respect to obstructive sleep apnoea, the recommendation from the AASM clinical practice guidelines was that actigraphy devices could be used in conjunction with a home sleep apnoea test to provide estimates of specific sleep parameters (namely total sleep time) (85). There is evidence to suggest that incorporating these estimates may provide slight improvements in the diagnostic accuracy of OSA (105, 106). More broadly, sleep monitoring technologies that assist oxygen saturation and/or breathing-related parameters are currently used to estimate OSA severity (107–109). Several such technologies are classed as medical devices and can therefore be used to screen for OSA symptoms. Randomised-controlled trials are needed to definitively demonstrate the effectiveness and cost-benefit of this approach.

In the future, there may be a greater role for sleep monitoring devices in the management of OSA. Multi-night measurement using accurate sleep and breathing monitors may give better indications of OSA severity than single-night sleep studies due to significant night-to-night variability in OSA severity. Multiple studies have found that the misdiagnosis rate for mild to moderate OSA can be as high as 20–50% when using these studies. This is supported by smaller studies with sample sizes up to 300 participants, which have reported misdiagnosis rates of up to 60% with single-night studies (110–115). Aside from misdiagnosis, night-to-night variability in OSA severity may also be a clinically meaningful feature associated with more severe adverse health outcomes (116, 117). If so, then sleep technologies would be necessary to assess for this feature in a feasible manner. Appropriately designed randomised-controlled trials are needed to test the clinical utility of simplified monitoring of OSA severity via sleep monitoring technologies and its potential benefits for supporting diagnosis and treatment compared to current diagnostic practices.

Sleep improvement for the general population

Sleep is a high-value biological process for ongoing monitoring and intervention where needed, as it has profound impacts on overall health and functioning. Given this importance, many individuals are currently using sleep monitoring technologies to track their sleep patterns over time. Several smartphone apps used to operate sleep monitoring devices provide personalised feedback to users based on their sleep data, often by comparing their sleep estimates to other users and/or population norms. Some apps also provide overall scores to users based on their sleep data, such as an overall rating of their prior night's sleep quality or a score to indicate their anticipated degree of next-day functioning. Such feedback, if presented appropriately, could translate into actions that improve an individual's sleep health. More research is needed to test whether this approach to providing feedback helps or hinders people's sleep and overall functioning.

One promising approach is to capitalise on the large and ever-growing number of people who are using devices capable of sleep monitoring to assess and support community-level sleep health education and improvement strategies. Strategies could include identifying individuals with poor sleep patterns, providing evidence-based guidance to adopt healthier sleep behaviours, administering efficacious therapies to improve sleep and functioning (e.g., digital CBT-I), and/or better connect individuals to sleep treatment services. With the widespread use of these technologies, the sleep medicine field has an excellent opportunity to address and improve sleep health on a massive scale. Such efforts will require strong collaborations with industry partners to refine these approaches and successfully deliver.

The bedroom of the future

There are numerous new techniques being developed for assessing sleep and circadian rhythms which will be available in the near future. Multiple sleep monitoring technologies are entering the healthcare market, which use sophisticated acquisition and signal processing approaches for EEG and other physiological signals to provide simplified sleep studies. There is also a definitive trend towards multi-sensor sleep monitoring devices for capturing more useful physiological and environmental signals to improve the accuracy of sleep estimation. More non-contact technologies are also under development, including infrared video and motion tracking sensors to monitor respiration, head posture, body posture, and to detect abnormal breathing (118). Furthermore, combinations of multiple sensors show excellent potential for continuous monitoring of patients in the intensive care unit setting and may help provide insight into the role of sleep disruption as a contributor to common adverse outcomes in this setting such as delirium (119). Advanced monitoring devices in conjunction with biomathematical modelling are also being developed to support circadian rhythm assessment (120). Electronic chips that can be implanted in body patches are being developed to enable the assessment of the cortisol rhythm through sweat (121). Emerging techniques for assessing both sleep and circadian rhythm include the use of machine learning to make better use of the existing rich neurophysiological information acquired from sleep technologies and better identify novel markers of sleep disruption. Inevitable advancements in acquisition systems and signal processing techniques will mean that objective sleep monitoring technologies will become more accurate, feasible and cost-effective for routine use as part of clinical care and general health promotion.

While it is difficult to predict exactly which and how sleep monitoring technologies will be implemented in bedrooms in the future (36), what is becoming increasingly apparent is that technology will likely be a critical feature for many individuals to manage their sleep health (122). These technologies and applications may include:

- Ongoing monitoring of sleep disorder symptom severity, integrated with the individual's clinical care, for automated analysis and adjustment of treatments as needed.
- Identification of early markers of disease onset and progression, to support an individual's overall health.
- Real-time and useful feedback regarding next-day performance and functioning, which may have important implications for work management in some occupations.
- Smart beds and mattresses to monitor and adjust body position as needed.
- Sophisticated systems to provide sufficient lighting at the appropriate times relative to the individual's sleep, circadian phase and desired wake-sleep schedule.
- Automated systems to maintain a suitable environment (i.e., temperature, light, noise) during the sleep period and support transitions to/from sleep period (i.e., wind-down and wake-up routines).

Overall, the bedroom of the future may be a space that is more connected and personalised to the needs and preferences of the individual.

Conclusion

This chapter reviewed common and emerging techniques for objectively assessing sleep and circadian rhythms being integrated into wearable and non-wearable sleep monitoring technologies. The accuracy of these technologies varies and is influenced by factors such as the sleep scoring algorithms and individual or environmental characteristics. While validation studies have largely been conducted in controlled settings with young, healthy individuals with good sleep, the accuracy of these technologies in the uncontrolled home environment with people who have poor sleep or sleep disorders is less clear. It is also unclear what level of accuracy is required for sleep monitoring devices to be suitable for specific applications. Some suggest that subjective judgement about the capabilities of the device should be used to determine its appropriateness, while others argue that proof-of-concept studies should be conducted to demonstrate feasibility and effectiveness for a specific purpose. We would argue that a combination of these approaches, followed by well-designed randomised-controlled trials, is necessary to thoroughly test sleep monitoring technologies for use in clinical practice.

Evidently, there are many current and future potential applications of sleep monitoring technology, particularly for aiding in the diagnosis and treatment of insomnia and sleep-disordered breathing. These devices can provide unique and clinically useful information, such as on insomnia phenotypes and multi-night variability of OSA severity. Considerable research is needed to develop and demonstrate the effectiveness and cost-benefit of using sleep monitoring technologies to support the clinical care of sleep disorders. Enabled by sophisticated monitoring and integration systems currently under development and/or likely possible in the 'bedroom of the future', there is considerable potential to develop and implement intervention strategies *en masse* in the general population. Thus, sleep monitoring technologies may enable the sleep medicine field to do the unprecedented: to intervene and improve sleep health on a global community level.

References

1 La Vaque T. The history of EEG Hans Berger: psychophysiologist. A historical vignette. J Neurother. 1999;3(2):1–9.
2 Fekedulegn D, Andrew ME, Shi M, Violanti JM, Knox S, Innes KE. Actigraphy-based assessment of sleep parameters. Ann Work Expo Health. 2020;64(4):350–67.
3 Sadeh A, Acebo C. The role of actigraphy in sleep medicine. Sleep Med Rev. 2002;6(2):113–24.
4 Ancoli-Israel S, Cole R, Alessi C, Chambers M, Moorcroft W, Pollak CP. The role of actigraphy in the study of sleep and circadian rhythms. Sleep. 2003;26(3):342–92.
5 Webster JB, Kripke DF, Messin S, Mullaney DJ, Wyborney G. An activity-based sleep monitor system for ambulatory use. Sleep. 1982;5(4):389–99.
6 Depner CM, Cheng PC, Devine JK, Khosla S, de Zambotti M, Robillard R, et al. Wearable technologies for developing sleep and circadian biomarkers: a summary of workshop discussions. Sleep. 2020;43(2):zsz254.
7 Kolla BP, Mansukhani S, Mansukhani MP. Consumer sleep tracking devices: a review of mechanisms, validity and utility. Expert Rev Med Devices. 2016;13(5):497–506.
8 Scott H, Lack L, Lovato N. A systematic review of the accuracy of sleep wearable devices for estimating sleep onset. Sleep Med Rev. 2019;49:101227.
9 Chinoy ED, Cuellar JA, Huwa KE, Jameson JT, Watson CH, Bessman SC, et al. Performance of seven consumer sleep-tracking devices compared with polysomnography. Sleep. 2021;44(5):zsaa291.
10 Biswas D, Simões-Capela N, Van Hoof C, Van Helleputte N. Heart rate estimation from wrist-worn photoplethysmography: a review. IEEE Sens J. 2019;19(16):6560–70.

11 Saquib N, Papon MTI, Ahmad I, Rahman A, editors. Measurement of heart rate using photoplethysmography. In: 2015 International Conference on Networking Systems and Security (NSysS). IEEE; 2015.

12 Versace F, Mozzato M, De Min Tona G, Cavallero C, Stegagno L. Heart rate variability during sleep as a function of the sleep cycle. Biol Psychol. 2003;63(2):149–62.

13 Tobaldini E, Nobili L, Strada S, Casali KR, Braghiroli A, Montano N. Heart rate variability in normal and pathological sleep. Front Physiol. 2013;4:294.

14 Korkalainen H, Aakko J, Duce B, Kainulainen S, Leino A, Nikkonen S, et al. Deep learning enables sleep staging from photoplethysmogram for patients with suspected sleep apnea. Sleep. 2020;43(11):zsaa098.

15 Davies HJ, Williams I, Peters NS, Mandic DP. In-ear spo2: a tool for wearable, unobtrusive monitoring of core blood oxygen saturation. Sensors. 2020;20(17):4879.

16 Jayawardhana M, de Chazal P, editors. Enhanced detection of sleep apnoea using heart-rate, respiration effort and oxygen saturation derived from a photoplethysmography sensor. In: 2017 39th Annual International Conference of the IEEE Engineering in Medicine and Biology Society (EMBC). IEEE; 2017.

17 Bazurto R, Hedayatipour A, editors. CMOS PPG sensor with correcting feedback for effects of skin pigmentation. In: 2022 29th IEEE International Conference on Electronics, Circuits and Systems (ICECS). IEEE; 2022.

18 Van Gastel M, Stuijk S, de Haan G. Motion robust remote-PPG in infrared. IEEE Trans Biomed Eng. 2015;62(5):1425–33.

19 Lack L, Gradisar M. Acute finger temperature changes preceding sleep onset over a 45-h period. J Sleep Res. 2002;11(4):275–82.

20 Gradisar M, Lack L, Wright H, Harris J, Brooks A. Do chronic primary insomniacs have impaired heat loss when attempting sleep? Am J Physiol Regul Integr Comp Physiol. 2006;290(4):R1115–21.

21 Gradisar M, Lack L. Relationships between the circadian rhythms of finger temperature, core temperature, sleep latency, and subjective sleepiness. J Biol Rhythms. 2004;19(2):157–63.

22 Raymann RJ, Swaab DF, Van Someren EJ. Skin temperature and sleep-onset latency: changes with age and insomnia. Physiol Behav. 2007;90(2–3):257–66.

23 Boe AJ, McGee Koch LL, O'Brien MK, Shawen N, Rogers JA, Lieber RL, et al. Automating sleep stage classification using wireless, wearable sensors. NPJ Digit Med. 2019;2:131.

24 Ortiz-Tudela E, Martinez-Nicolas A, Campos M, Rol MA, Madrid JA. A new integrated variable based on thermometry, actimetry and body position (TAP) to evaluate circadian system status in humans. PLoS Comput Biol. 2010;6(11):e1000996.

25 Herborn KA, Graves JL, Jerem P, Evans NP, Nager R, McCafferty DJ, et al. Skin temperature reveals the intensity of acute stress. Physiol Behav. 2015;152:225–30.

26 Mekjavic IB, Eiken O. Contribution of thermal and nonthermal factors to the regulation of body temperature in humans. J Appl Physiol. 2006;100(6):2065–72.

27 Dement W, Kleitman N. Cyclic variations in EEG during sleep and their relation to eye movements, body motility, and dreaming. EEG Clin Neurophysiol. 1957;9(4):673–90.

28 Berry RB, Brooks R, Gamaldo CE, Harding SM, Lloyd RM, Marcus CL, et al. The AASM Manual for the Scoring of Sleep and Associated Events: Rules, Terminology and Technical Specifications. Darien, IL: American Academy of Sleep Medicine; 2015.

29 Zhang Z, Guan CT, Chan TE, Yu JH, Ng AK, Zhang HH, et al. Automatic sleep onset detection using single EEG sensor. In: 2014 36th Annual International Conference of the Ieee Engineering in Medicine and Biology Society. IEEE Engineering in Medicine and Biology Society Conference Proceedings; 2014. p. 2265–8.

30 Lucey BP, McLeland JS, Toedebusch CD, Boyd J, Morris JC, Landsness EC, et al. Comparison of a single-channel EEG sleep study to polysomnography. J Sleep Res. 2016;25(6):625–35.

31 Arnal PJ, Thorey V, Debellemaniere E, Ballard ME, Bou Hernandez A, Guillot A, et al. The dreem headband compared to polysomnography for EEG signal acquisition and sleep staging. Sleep. 2020;43(11):zsaa097.

32 Palotti J, Mall R, Aupetit M, Rueschman M, Singh M, Sathyanarayana A, et al. Benchmark on a large cohort for sleep-wake classification with machine learning techniques. NPJ Digit Med. 2019;2:50.

33 Abou Jaoude M, Sun H, Pellerin KR, Pavlova M, Sarkis RA, Cash SS, et al. Expert-level automated sleep staging of long-term scalp EEG recordings using deep learning. Sleep. 2020;43(11):zsaa112.

34 Prerau MJ, Hartnack KE, Obregon-Henao G, Sampson A, Merlino M, Gannon K, et al. Tracking the sleep onset process: an empirical model of behavioral and physiological dynamics. PLoS Comput Biol. 2014;10(10):e1003866.

35 Scott H, Lechat B, Lovato N, Lack L. Correspondence between physiological and behavioural responses to vibratory stimuli during the sleep onset period: a quantitative electroencephalography analysis. J Sleep Res. 2021;30(4):e13232.

36 Lechat B, Scott H, Naik G, Hansen K, Nguyen DP, Vakulin A, et al. New and emerging approaches to better define sleep disruption and its consequences. Front Neurosci. 2021;15.

37 Tizard B. Evoked changes in EEG and electrodermal activity during the waking and sleeping states. Electroencephalogr Clin Neurophysiol. 1966;20(2):122–8.

38 Sano A, Picard RW, editors. Toward a taxonomy of autonomic sleep patterns with electrodermal activity. In: 2011 Annual International Conference of the IEEE. Boston, US: IEEE; 2011.

39 Sano A, Picard RW, Stickgold R. Quantitative analysis of wrist electrodermal activity during sleep. Int J Psychophysiol. 2014;94(3):382–9.

40 Herlan A, Ottenbacher J, Schneider J, Riemann D, Feige B. Electrodermal activity patterns in sleep stages and their utility for sleep versus wake classification. J Sleep Res. 2018:e12694.

41 Martin WB, Johnson LC, Viglione SS, Naitoh P, Joseph RD, Moses JD. Pattern recognition of EEG-EOG as a technique for all-night sleep stage scoring. Electroencephalogr Clin Neurophysiol. 1972;32:417–27.

42 Fan J, Sun C, Long M, Chen C, Chen W. EOGNET: a novel deep learning model for sleep stage classification based on single-channel EOG signal. Front Neurosci. 2021;15.

43 Myllymaa S, Muraja-Murro A, Westeren-Punnonen S, Hukkanen T, Lappalainen R, Mervaala E, et al. Assessment of the suitability of using a forehead EEG electrode set and chin EMG electrodes for sleep staging in polysomnography. J Sleep Res. 2016;25(6):636–45.

44 Yachida W, Arima T, Castrillon EE, Baad-Hansen L, Ohata N, Svensson P. Diagnostic validity of self-reported measures of sleep bruxism using an ambulatory single-channel EMG device. J Prosthodont Res. 2016;60(4):250–7.

45 Martinot J-B, Le-Dong N-N, Letesson C, Cuthbert V, Gozal D, Pepin J-L. Design and validation of a new technology for home-based screening of sleep apnea syndrome. In: C109 SRN: Innovative Ways to Assess SDB and Predict Outcomes to Current Therapies. American Thoracic Society; 2020. p. A6160-A.

46 Martinot J, Le-Dong N, Cuthbert V, Denison S, Gozal D, Pepin J. 0792 Mandibular movement monitoring with artificial intelligence analysis for the diagnosis of sleep bruxism. Sleep. 2020;43(Supplement_1):A301–2.

47 Pion-Massicotte J, Godbout R, Savard P, Roy JF. Development and validation of an algorithm for the study of sleep using a biometric shirt in young healthy adults. J Sleep Res. 2019;28(2):e12667.

48 Karlen W, Mattiussi C, Floreano D. Sleep and wake classification with ECG and respiratory effort signals. IEEE Trans Biomed Circuits Syst. 2009;3(2):71–8.

49 Sun H, Ganglberger W, Panneerselvam E, Leone MJ, Quadri SA, Goparaju B, et al. Sleep staging from electrocardiography and respiration with deep learning. Sleep. 2020;43(7):zsz306.
50 Rahman T, Page R, Page C, Bonnefoy J-R, Cox T, Shaffer TH. pneuRIPTM: a novel respiratory inductance plethysmography monitor. J Med Device. 2017;11(1).
51 Gubner RS, Rodstein M, Ungerleider HE. Ballistocardiography: an appraisal of technic, physiologic principles, and clinical value. Circulation. 1953;7(2):268–86.
52 Giovangrandi L, Inan OT, Wiard RM, Etemadi M, Kovacs GT, editors. Ballistocardiography – a method worth revisiting. In: 2011 Annual International Conference of the IEEE Engineering in Medicine and Biology Society. IEEE; 2011.
53 Park KS, Hwang SH, Yoon HN, Lee WK, editors. Ballistocardiography for nonintrusive sleep structure estimation. In: 2014 36th Annual International Conference of the IEEE Engineering in Medicine and Biology Society. IEEE; 2014.
54 Huysmans D, Borzée P, Testelmans D, Buyse B, Willemen T, Van Huffel S, et al. Evaluation of a commercial ballistocardiography sensor for sleep apnea screening and sleep monitoring. Sensors. 2019;19(9):2133.
55 Edouard P, Campo D, Bartet P, Yang RY, Bruyneel M, Roisman G, et al. Validation of the withings sleep analyzer, an under-the-mattress device for the detection of moderate-severe sleep apnea syndrome. J Clin Sleep Med. 2021;17(6):1217–27.
56 Chen VC, Li F, Ho S-S, Wechsler H. Micro-Doppler effect in radar: phenomenon, model, and simulation study. IEEE Trans Aerosp Electron Syst. 2006;42(1):2–21.
57 Heglum HSA, Kallestad H, Vethe D, Langsrud K, Sand R, Engstrom M. Distinguishing sleep from wake with a radar sensor a contact-free real-time sleep monitor. Sleep. 2021;44(8):zsab060.
58 Zaffaroni A, Coffey S, Dodd S, Kilroy H, Lyon G, O'Rourke D, et al., editors. Sleep staging monitoring based on sonar smartphone technology. In: 2019 41st Annual International Conference of the IEEE Engineering in Medicine and Biology Society (EMBC). IEEE; 2019.
59 Lyon G, Tiron R, Zaffaroni A, Osman A, Kilroy H, Lederer K, et al., editors. Detection of sleep apnea using sonar smartphone technology. In: 2019 41st Annual International Conference of the IEEE Engineering in Medicine and Biology Society (EMBC). IEEE; 2019.
60 de Zambotti M, Cellini N, Menghini L, Sarlo M, Baker FC. Sensors capabilities, performance, and use of consumer sleep technology. Sleep Med Clin. 2020;15(1):1–30.
61 de Zambotti M, Cellini N, Goldstone A, Colrain IM, Baker FC. Wearable sleep technology in clinical and research settings. Med Sci Sports Exerc. 2019;51(7):1538–57.
62 Stone JE, McGlashan EM, Quin N, Skinner K, Stephenson JJ, Cain SW, et al. The role of light sensitivity and intrinsic circadian period in predicting individual circadian timing. J Biol Rhythms. 2020;35(6):628–40.
63 Tekieh T, Lockley SW, Robinson PA, McCloskey S, Zobaer MS, Postnova S. Modeling melanopsin-mediated effects of light on circadian phase, melatonin suppression, and subjective sleepiness. J Pineal Res. 2020:e12681.
64 Ong AA, Gillespie MB. Overview of smartphone applications for sleep analysis. World J Otorhinolaryngol Head Neck Surg. 2016;2(1):45–9.
65 Bianchi MT. Consumer sleep apps: when it comes to the big picture, it's all about the frame. J Clin Sleep Med. 2015;11(7):695–6.
66 Fino E, Plazzi G, Filardi M, Marzocchi M, Pizza F, Vandi S, et al. (Not so) smart sleep tracking through the phone: findings from a polysomnography study testing the reliability of four sleep applications. J Sleep Res. 2019:e12935.
67 Nakano H, Hirayama K, Sadamitsu Y, Toshimitsu A, Fujita H, Shin S, et al. Monitoring sound to quantify snoring and sleep apnea severity using a smartphone: proof of concept. J Clin Sleep Med. 2014;10(1):73–8.

68 Castillo Y, Cámara MA, Blanco-Almazán D, Jané R, editors. Characterization of microphones for snoring and breathing events analysis in mHealth. In: 2017 39th Annual International Conference of the IEEE Engineering in Medicine and Biology Society (EMBC). IEEE; 2017.

69 Basner M, McGuire S. WHO environmental noise guidelines for the European region: a systematic review on environmental noise and effects on sleep. Int J Environ Res Public Health. 2018;15(3):519.

70 Basner M, Babisch W, Davis A, Brink M, Clark C, Janssen S, et al. Auditory and non-auditory effects of noise on health. Lancet. 2014;383(9925):1325–32.

71 Muzet A. Environmental noise, sleep and health. Sleep Med Rev. 2007;11(2):135–42.

72 Rosenberger ME, Buman MP, Haskell WL, McConnell MV, Carstensen LL. Twenty-four hours of sleep, sedentary behavior, and physical activity with nine wearable devices. Med Sci Sports Exerc. 2016;48(3):457–65.

73 Cook JD, Prairie ML, Plante DT. Utility of the Fitbit Flex to evaluate sleep in major depressive disorder: a comparison against polysomnography and wrist-worn actigraphy. J Affect Disord. 2017;217:299–305.

74 de Zambotti M, Goldstone A, Claudatos S, Colrain IM, Baker FC. A validation study of Fitbit charge 2 compared with polysomnography in adults. Chronobiol Int. 2018;35(4):465–76.

75 Kang SG, Kang JM, Ko KP, Park SC, Mariani S, Weng J. Validity of a commercial wearable sleep tracker in adult insomnia disorder patients and good sleepers. J Psychosom Res. 2017;97:38–44.

76 Kahawage P, Jumabhoy R, Hamill K, de Zambotti M, Drummond SPA. Validity, potential clinical utility, and comparison of consumer and research-grade activity trackers in insomnia disorder I: In-lab validation against polysomnography. J Sleep Res. 2019:e12931.

77 Meltzer LJ, Walsh CM, Traylor J, Westin AM. Direct comparison of two new actigraphs and polysomnography in children and adolescents. Sleep. 2012;35(1):159–66.

78 Miller DJ, Sargent C, Roach GD. A validation of six wearable devices for estimating sleep, heart rate and heart rate variability in healthy adults. Sensors (Basel). 2022;22(16).

79 Shambroom JR, Fabregas SE, Johnstone J. Validation of an automated wireless system to monitor sleep in healthy adults. J Sleep Res. 2012;21(2):221–30.

80 Griessenberger H, Heib DPJ, Kunz AB, Hoedlmoser K, Schabus M. Assessment of a wireless headband for automatic sleep scoring. Sleep Breath. 2013;17(2):747–52.

81 Quante M, Kaplan ER, Cailler M, Rueschman M, Wang R, Weng J, et al. Actigraphy-based sleep estimation in adolescents and adults: a comparison with polysomnography using two scoring algorithms. Nat Sci Sleep. 2018;10:13–20.

82 Danzig R, Wang M, Shah A, Trotti LM. The wrist is not the brain: estimation of sleep by clinical and consumer wearable actigraphy devices is impacted by multiple patient- and device-specific factors. J Sleep Res. 2019:e12926.

83 Rosenberg RS, Van Hout S. The American academy of sleep medicine inter-scorer reliability program: sleep stage scoring. J Clin Sleep Med. 2013;9(1):81–7.

84 Baron KG, Duffecy J, Berendsen MA, Cheung IN, Lattie E, Manalo NC. Feeling validated yet? A scoping review of the use of consumer-targeted wearable and mobile technology to measure and improve sleep. Sleep Med Rev. 2018;40:151–9.

85 Smith MT, Mccrae CS, Cheung J, Martin JL, Harrod CG, Heald JL, et al. Use of actigraphy for the evaluation of sleep disorders and circadian rhythm sleep-wake disorders: an American academy of sleep medicine clinical practice guideline. J Clin Sleep Med. 2018;14(7):1231–7.

86 Smith MT, McCrae CS, Cheung J, Martin JL, Harrod CG, Heald JL, et al. Use of actigraphy for the evaluation of sleep disorders and circadian rhythm sleep-wake disorders: an American academy of sleep medicine systematic review, meta-analysis, and GRADE assessment. J Clin Sleep Med. 2018;14(7):1209–30.

87 Fernandez-Mendoza J. The insomnia with short sleep duration phenotype: an update on it's importance for health and prevention. Curr Opin Psychiatry. 2017;30(1).

88 Vgontzas AN, Fernandez-Mendoza J, Liao D, Bixler EO. Insomnia with objective short sleep duration: the most biologically severe phenotype of the disorder. Sleep Med Rev. 2013;17(4):241–54.

89 Fernandez-Mendoza J, Vgontzas AN, Liao D, Shaffer ML, Vela-Bueno A, Basta M, et al. Insomnia with objective short sleep duration and incident hypertension. Hypertension. 2012;60(4):929–35.

90 Fernandez-Mendoza J, Shea S, Vgontzas AN, Calhoun SL, Liao D, Bixler EO. Insomnia and incident depression: role of objective sleep duration and natural history. J Sleep Res. 2015;24(4):390–8.

91 Vgontzas AN, Liao D, Pejovic S, Calhoun S, Karataraki M, Basta M, et al. Insomnia with short sleep duration and mortality: the penn state cohort. Sleep. 2010;33(9):1159–64.

92 Bensen-Boakes D-B, Lovato N, Meaklim H, Bei B, Scott H. "Sleep-wake state discrepancy": toward a common understanding and standardized nomenclature. Sleep. 2022;45(10):zsac187.

93 Edinger JD, Krystal AD. Subtyping primary insomnia: is sleep state misperception a distinct clinical entity? Sleep Med Rev. 2003;7(3):203–14.

94 Harris J, Lack L, Kemp K, Wright H, Bootzin R. A randomized controlled trial of intensive sleep retraining (ISR): a brief conditioning treatment for chronic insomnia. Sleep. 2012;35(1):49–60.

95 Harris J, Lack L, Wright H, Gradisar M, Brooks A. Intensive sleep retraining treatment for chronic primary insomnia: a preliminary investigation. J Sleep Res. 2007;16(3):276–84.

96 Lack L, Scott H, Micic G, Lovato N. Intensive sleep re-training: from bench to bedside. Brain Sci. 2017;7(4).

97 Scott H, Lovato N, Lack L. The development and accuracy of the THIM wearable device for estimating sleep and wakefulness. Nat Sci Sleep. 2021;13:39.

98 Scott H, Whitelaw A, Canty A, Lovato N, Lack L. The accuracy of the THIM wearable device for estimating sleep onset latency. J Clin Sleep Med. 2021:jcsm.9070.

99 Erten Uyumaz B, Feijs L, Hu J. A review of digital cognitive behavioral therapy for insomnia (CBT-I Apps): are they designed for engagement? Int J Environ Res Public Health. 2021;18(6).

100 Hasan F, Tu YK, Yang CM, James Gordon C, Wu D, Lee HC, et al. Comparative efficacy of digital cognitive behavioral therapy for insomnia: a systematic review and network meta-analysis. Sleep Med Rev. 2022;61:101567.

101 Soh HL, Ho RC, Ho CS, Tam WW. Efficacy of digital cognitive behavioural therapy for insomnia: a meta-analysis of randomised controlled trials. Sleep Med. 2020;75:315–25.

102 Aji M, Glozier N, Bartlett D, Peters D, Calvo RA, Zheng Y, et al. A feasibility study of a mobile app to treat insomnia. Transl Behav Med. 2021;11(2):604–12.

103 Baron KG, Abbott S, Jao N, Manalo N, Mullen R. Orthosomnia: are some patients taking the quantified self too far? J Clin Sleep Med. 2017;13(2):351–4.

104 Baron KG. CBT-I for patients with orthosomnia. In: Adapting Cognitive Behavioral Therapy for Insomnia. Netherlands: Elsevier; 2022. p. 135–45.

105 Choi SJ, Kang M, Sung MJ, Joo EY. Discordant sleep parameters among actigraphy, polysomnography, and perceived sleep in patients with sleep-disordered breathing in comparison with patients with chronic insomnia disorder. Sleep Breath. 2017;21(4):837–43.

106 Sharif MM, Bahammam AS. Sleep estimation using BodyMedia's SenseWear armband in patients with obstructive sleep apnea. Ann Thorac Med. 2013;8(1):53–7.

107 Choi JH, Kim EJ, Kim YS, Choi J, Kim TH, Kwon SY, et al. Validation study of portable device for the diagnosis of obstructive sleep apnea according to the new AASM scoring criteria: watch-PAT 100. Acta Otolaryngol. 2010;130(7):838–43.

108 Gu W, Leung L, Kwok KC, Wu I-C, Folz RJ, Chiang AA. Belun ring platform: a novel home sleep apnea testing system for assessment of obstructive sleep apnea. J Clin Sleep Med. 2020;16(9):1611–17.

109 Ummel JD, Hoilett OS, Walters BD, Bluhm ND, Pickering AS, Wilson DA, et al. editors. Kick ring LL: a multi-sensor ring capturing respiration, electrocardiogram, oxygen saturation, and skin temperature. In: 2020 42nd Annual International Conference of the IEEE Engineering in Medicine & Biology Society (EMBC). IEEE; 2020.

110 Punjabi NM, Patil S, Crainiceanu C, Aurora RN. Variability and misclassification of sleep apnea severity based on multi-night testing. Chest. 2020;158(1):365–73.

111 Joosten SA, O'Donoghue FJ, Rochford PD, Barnes M, Hamza K, Churchward TJ, et al. Night-to-night repeatability of supine-related obstructive sleep apnea. Ann Am Thorac Soc. 2014;11(5):761–9.

112 Skiba V, Goldstein C, Schotland H. Night-to-night variability in sleep disordered breathing and the utility of esophageal pressure monitoring in suspected obstructive sleep apnea. J Clin Sleep Med. 2015;11(6):597–602.

113 Stoberl AS, Schwarz EI, Haile SR, Turnbull CD, Rossi VA, Stradling JR, et al. Night-to-night variability of obstructive sleep apnea. J Sleep Res. 2017;26(6):782–8.

114 Anitua E, Duran-Cantolla J, Almeida GZ, Alkhraisat MH. Predicting the night-to-night variability in the severity of obstructive sleep apnea: the case of the standard error of measurement. Sleep Sci. 2019;12(2):72–8.

115 Tschopp S, Wimmer W, Caversaccio M, Borner U, Tschopp K. Night-to-night variability in obstructive sleep apnea using peripheral arterial tonometry: a case for multiple night testing. J Clin Sleep Med. 2021;17(9):1751–8.

116 Linz D, Brooks AG, Elliott AD, Nalliah CJ, Hendriks JML, Middeldorp ME, et al. Variability of sleep apnea severity and risk of atrial fibrillation: the VARIOSA-AF study. JACC Clin Electrophysiol. 2019;5(6):692–701.

117 Lechat B, Loffler K, Reynolds A, Naik G, Aishah A, Scott H, et al. Association between night-to-night variability in obstructive sleep apnea severity with hypertension risk and blood pressure variability. Am J Respir Crit Care Med. 2022;205(205):A3653–569.

118 Deng F, Dong J, Wang X, Fang Y, Liu Y, Yu Z, et al. Design and implementation of a noncontact sleep monitoring system using infrared cameras and motion sensor. IEEE Trans Instrum Meas. 2018;67(7):1555–63.

119 Davoudi A, Malhotra KR, Shickel B, Siegel S, Williams S, Ruppert M, et al. Intelligent ICU for autonomous patient monitoring using pervasive sensing and deep learning. Sci Rep. 2019;9(1):1–13.

120 Reid KJ. Assessment of circadian rhythms. Neurol Clin. 2019;37(3):505–26.

121 Upasham S, Prasad S. SLOCK (sensor for circadian clock): passive sweat-based chronobiology tracker. Lab Chip. 2020;20(11):1947–60.

122 Perez-Pozuelo I, Zhai B, Palotti J, Mall R, Aupetit M, Garcia-Gomez JM, et al. The future of sleep health: a data-driven revolution in sleep science and medicine. NPJ Digit Med. 2020;3(1):1–15.

INDEX